LAGRANGIAN MECHANICS

Edited by **Hüseyin Canbolat**

Lagrangian Mechanics
http://dx.doi.org/10.5772/63168
Edited by Hüseyin Canbolat

Contributors

Uziel Sandler, Lev Tsitolovsky, Leo Butler, Samuel Frimpong, Muhammad Azeem Raza, Yang Liu, Efrain Alcorta-Garcia, Cesar Martinez Torres, Luis Humberto Rodriguez Alfaro, Gerardo Romero, David Lara, Yury Vasilyevich Yanilkin, Alvaro Restuccia Nuñez, Adrian Sotomayor

Published by InTech

Janeza Trdine 9, 51000 Rijeka, Croatia

Publishing Process Manager Romina Skomersic
Technical Editor SPi Global
Cover InTech Design team

First published May, 2017
Printed in Croatia
Legal deposit, Croatia: National and University Library in Zagreb

Additional hard copies can be obtained from orders@intechopen.com

Lagrangian Mechanics, Edited by Hüseyin Canbolat
p. cm.
Print ISBN 978-953-51-3131-1
Online ISBN 978-953-51-3132-8

For EU product safety concerns:
IN TECH d.o.o., Prolaz Marije Krucifikse Kozulić 3, 51000 Rijeka, Croatia,
info@intechopen.com or visit our website at intechopen.com.

Meet the editor

Dr. Hüseyin Canbolat was born in Adana in 1966. He received his BS and MS degrees from the Middle East Technical University, Ankara, Turkey, in 1989 and 1993, respectively, and PhD degree in Electrical Engineering from Clemson University, Clemson, SC, USA, in 1997. After his PhD studies, he joined the Department of Electrical and Electronic Engineering at Mersin University in 1998. Since 2012, he has been working with the Department of Electrical and Electronic Engineering, Ankara Yildirim Beyazit University, Ankara, Turkey. His research interests include control systems with applications to robotic and mechatronic systems, MEMS, energy systems, measurement, and instrumentation. He is a senior member of IEEE.

Contents

Preface

Lagrangian mechanics is widely used in several areas of research and technology. It is simply a reformulation of the classical mechanics by the mathematician and astronomer Joseph-Louis Lagrange in 1788. The approach formulates the physical phenomena through a function called Lagrangian, which is a function of generalized coordinates and contains the dynamics of the system through the derivatives.

Lagrangian mechanics is good for systems with conservative forces. If the dissipative forces are included, these forces should be separated into potential and nonpotential forces. This formulation gives a set of modified Euler-Lagrange equations. The user can choose generalized coordinates such as the symmetries in the system or the geometry of the constraints. This may simplify the solutions for the motion of the system.

Lagrangian mechanics is also important for its role in deep understanding of physics besides its broad applications. It is applicable to most of the fundamental theoretical physics, such as quantum mechanics and relativity theory, even though Lagrange considered only the classical mechanics in *Mecanique Analytique*. Hamilton's principle is closely related to Lagrangian mechanics, since it can be employed in the derivation of Lagrangian equations.

Lagrangian mechanics can also be applied to other systems. An example of these systems is the coupled electric circuit including inductive and capacitive components. Lagrangian mechanics is a good alternative in solving mechanical problems in physics and engineering, especially when Newton's formulation of classical mechanics is not convenient. It can also be used in the optimization problems of dynamic systems.

In this book, the section authors provide state-of-the-art research studies on Lagrangian mechanics. Hopefully, the researchers will benefit from the book in their studies. It is probable that the presented studies may lead the researchers to develop new ideas in conducting their research.

Dr. Hüseyin Canbolat
Yildirim Beyazit University,
Ankara, Turkey

Section 1

Theory

Chapter 1

Singular Lagrangians and Its Corresponding Hamiltonian Structures

Alvaro Restuccia and Adrián Sotomayor

Additional information is available at the end of the chapter

http://dx.doi.org/10.5772/66146

Abstract

We present a general procedure to obtain the Lagrangian and associated Hamiltonian structure for integrable systems of the Helmholtz type. We present the analysis for coupled Korteweg-de Vries systems that are extensions of the Korteweg-de Vries equation. Starting with the system of partial differential equations it is possible to follow the Helmholtz approach to construct one or more Lagrangians whose stationary points coincide with the original system. All the Lagrangians are singular. Following the Dirac approach, we obtain all the constraints of the formulation and construct the Poisson bracket on the physical phase space via the Dirac bracket. We show compatibility of some of these Poisson structures. We obtain the Gardner ε-deformation of these systems and construct a master Lagrangian which describe the coupled systems in the weak ε-limit and its modified version in the strong ε-limit.

Keywords: integrable systems, conservation laws, partial differential equations, rings and algebras

1. Introduction

The Lagrangian mechanics has a wide range of applications from classical mechanics to quantum field theory. There are two main reasons to introduce a Lagrangian in order to describe a physical model. Its stationary points, defined in terms of functional derivatives, provide the classical equations of motion or classical field equations governing the evolution of the physical system while the action functional constructed from the Lagrangian provides the path integral approach to quantum mechanics and quantum field theories. In this chapter, we analyze several aspects of singular Lagrangians, which are relevant in various areas of physics. They are essential in the description of the fundamental forces in nature and in the analysis of integrable systems. In this chapter, we consider recent applications of singular Lagrangians in the area of completely integrable systems.

The analysis of integrable systems, in particular the Korteweg-de Vries equation and extensions of it [1–16], have provided a lot of interesting results from both mathematical and physical points of view.

Besides the physical applications of coupled KdV systems at low energies [17–19], one of the Poisson structures of the KdV equation is related to the Virasoro algebra with central terms. The latest is a fundamental symmetry of string theory, a proposal for a consistent quantum gravity theory.

In this chapter, we discuss a general approach based on the Helmholtz procedure to obtain a Lagrangian formulation and the Hamiltonian structure, starting from the system of time evolution partial differential equations describing the coupled KdV systems. Once the Lagrangian, whose stationary points corresponds to the integrable equations, has been obtained we follow the Dirac approach to constrained systems [20] to obtain the complete set of constraints and the Hamiltonian structure of the system. We discuss the existence of more than one Poisson structures associated with the integrable systems. Some of them are compatible Poisson structures and define a pencil of Poisson structures. We also discuss duality relations among the integrable systems we consider. The extensions of the KdV equation include a parametric coupled KdV system [21, 22], which we discuss in Section 3. In Section 8, we present a coupled KdV system arising from the breaking of a $N = 1$ supersymmetric model [15]. In Section 11, we discuss an extension of the KdV equation where the fields are valued on the octonion algebra and the product in the equation is the product on the octonion algebra [23]. This system has a supersymmetric extension which may be directly related to a model of the $D = 11$ supermembrane theory, a relevant sector of M–theory. The latest is a proposal of unification of all fundamental forces at very high energy.

2. The Dirac procedure for constrained systems

The Dirac approach for constrained systems [20] is a fundamental tool in the analysis of classical and quantum aspects of a physical theory. From a classical point of view, it provides a precise formulation of the initial valued problem for a time evolution system of partial differential equations. The initial data for the initial valued problem, given in terms of a constrained submanifold of a phase space, defines the physical phase space provided with the corresponding Poisson structure which gives rise to the canonical quantization of the system. In field theory, the starting point is a Lagrangian formulation. Its stationary points determine the classical field equations, generically a time evolution system of partial differential equations. From the Lagrangian density $\mathcal{L}$, one defines the conjugate momenta $p_i, i = 1, \ldots, N,$ associated with the original independent fields $q_i, i = 1, \ldots, N,$ defining the Lagrangian:

$$p_i = \frac{\partial \mathcal{L}}{\partial \dot{q}_i}. \tag{1}$$

$\mathcal{L}$ is assumed to be a function of $\dot{q}_i$ and a finite number of spatial derivatives.

http://dx.doi.org/10.5772/66146

If the Hessian matrix $\frac{\partial^2 \mathcal{L}}{\partial \dot{q}_i \partial \dot{q}_j}$ is singular we cannot express, from the above equation defining the conjugate momenta, all the $\dot{q}_i$ velocities in terms of the conjugate momenta.

The system presents then constraints on the phase space defined by the conjugate pairs $(q_i, p_i), i = 1, \ldots, N$. The phase space is provided with a Poisson structure given by

$$\{q_i, \ p_j\} = \delta_{ij}, \quad \{q_i, q_j\} = \{p_i, p_j\} = 0. \tag{2}$$

In general, it is a difficult task to disentangle all the constraints on the phase space associated with a given Lagrangian. The Dirac approach provides a systematic way to obtain all the constraints on phase space. Moreover, it determines the Lagrange multipliers associated with the constraints (eventually after a gauge fixing procedure) in a way that if the constraints are satisfied initially then the Hamilton equations ensure that they are satisfied at any time. In this sense, it provides a precise formulation of the initial value problem, the initial data is given by the set of (q_i, p_i) conjugate pairs satisfying the constraints on phase space. The Hamilton equations then provide the time evolution of the system. This constrained initial data, with its associated Poisson structure (also obtained from the Dirac construction) provides the fundamental structure to define the canonical quantization of the original Lagrangian.

From the equation defining momenta one obtains, in the case of singular Lagrangian, a set of constraints $\phi_M(q, p) = 0$, where the argument is a shorthand notation for p, q and their derivatives with respect to the spatial coordinates x_a, $a = 1, \ldots, k$.

Also, by performing a Legendre transformation one gets a Hamiltonian $H_0 = \int_{-\infty}^{+\infty} dx \mathcal{H}_0$, where the Hamilton density is given by

$$\mathcal{H}_0 = \sum_i p_i \dot{q}_i - \mathcal{L}, \tag{3}$$

where $\mathcal{L}$ is the Lagrangian density. Then, we obtain a new Hamiltonian $H = \int_{-\infty}^{+\infty} dx \mathcal{H}$ with a density $\mathcal{H} = \mathcal{H}_0 + \lambda_M \phi_M$. The conservation of the constraints, which have to be satisfied at any time, yields

$$\dot{\phi}_M = \{\phi_M, H\} = 0. \tag{4}$$

$\{\phi_M, H\} = 0$ may (i) be identically satisfied on the constrained surface $\phi_m = 0$,

(ii) determine Lagrange multipliers, or

(iii) give new constraints.

In Case (i) or (ii), the procedure ends; in Case (iii), the iteration follows exactly in the same way. At some step, the procedure ends, assuming that there is a finite of physical degrees of freedom describing the dynamics of the original Lagrangian. In the procedure, a set of Lagrange multipliers may be determined and others may not. The constraints associated with the ones that have been determined are called second class constraints, the other constraints for which the Lagrange multipliers are not determined are related to first class constraints. The first class constraints are the generators of a gauge symmetry on the time evolution system of partial differential equations. A difficult situation may occur in field theory when there is a combination of first and second class constraints. In order to separate them, one may have to invert some matrix involving fields of the formulation which may render dangerous non-localities in the final formulation.

All physical theories of the known fundamental forces in nature are formulated in terms of Lagrangians with gauge symmetries. All of them have first class constraints in their canonical formulation. In addition, they may also have second class constraints. In the analysis of field theories which are completely integrable systems like the ones we will discuss in this chapter only second class constraint appear. In this case, there are short cut procedures to simplify the Dirac procedure. However, the richness of the Dirac approach is that from its formulation one can extrapolate gauge systems which under a gauge fixing procedure reduce to the given system with second class constraints only. This is one of the main motivations of this chapter, to establish the Lagrangian and Hamiltonian structure for coupled KdV systems, which may allow the construction of gauge systems which are completely integrable.

In the case in which the constrained system has second class constraints, Dirac introduced the Poisson structure on the constrained submanifold in phase space. It determines the "physical" phase space with its Poisson bracket structure given by the Dirac bracket. They are defined in terms of the original Poisson bracket {, } on the full phase space by:

$$\{F, G\}_{DB} = -\{F, \phi_M\}\{\phi_M, \phi_N\}^{-1}\{\phi_N, G\} \tag{5}$$

where $\{\phi_M, \phi_N\}^{-1}$ is the inverse of the matrix $\{\phi_M, \phi_N\}$ which, in the case where $\phi_M = 0$ are second class constraints, is always of full rank.

The difficulty in field theory occurs when the matrix $\{\phi_M, \phi_N\}$ depends on the fields describing the theory and its inverse may lead to nonlocalities in the formulation. In our applications, those difficulties will not be present.

The Dirac bracket of a second class constraint with any other observable is zero. Consequently, the time conservation of the second class constraints is assured by the construction. For the same reason, there is no ambiguity on which Hamiltonian is used in determining the time evolution of observables.

3. A parametric coupled KdV system

A very interesting and well-known integrable system is the Korteweg-de Vries (KdV) equation. It arises from a variational principle of a singular Lagrangian. In what follows, we consider an extension of it. A coupled KdV system formulated in terms of two real differentiable functions $u(x, t)$ and $v(x, t)$ given by the following partial differential equations [21]:

$$u_t + uu_x + u_{xxx} + \lambda vv_x = 0 \tag{6}$$

$$v_t + u_x v + v_x u + v_{xxx} = 0 \tag{7}$$

where λ is a real parameter.

When discussing conserved quantities, we will assume that u and v belong to the real Schwartz space defined by

$$C_{\downarrow}^{\infty} = \left\{ w \in C^{\infty}(\mathbb{R}) / \lim_{x \to \pm\infty} x^p \frac{\partial^q}{\partial x^q} w = 0; p, q \geq 0 \right\} \tag{8}$$

When $\lambda = +1$ the system is equivalent to two decoupled KdV equations. When $\lambda = -1$ the system is equivalent to a KdV equation valued on the complex algebra. By a redefinition of v given by $v \to \frac{v}{\sqrt{|\lambda|}}$ the system for $\lambda > 0$ reduces to the $\lambda = +1$ case and the system for $\lambda < 0$ reduces to the $\lambda = -1$ case. The case $\lambda = 0$ is an independent integrable system.

The system (6) and (7) for $\lambda = -1$ describes a two-layer liquid model studied in references [17–19]. It is a very interesting evolution system. It is known to have solutions developing singularities on a finite time [24]. Also, a class of solitonic solutions was reported in [25] through the Hirota approach [26] and in [27] via a Bäcklund transformation in the sense of Wahlquist and Estabrook (WE) [28].

The system (6) and (7) for $\lambda = 0$ correspond to the ninth Hirota-Satsuma [6] coupled KdV system given in Ref. [29] (for the particular value of $k = 0$) (see also [30]) and is also included in the interesting study that relates integrable hierarchies with polynomial Lie algebras [31].

4. The Lagrangian associated with the parametric coupled KdV system

In this section, we obtain the Lagrangian and associated Hamiltonian structure of the coupled KdV system. We present the main results in Ref. [22].

The Lagrangian construction requires the introduction of the Casimir potentials w and y given by

$$\begin{aligned} u(x,t) &= w_x(x,t) \\ v(x,t) &= y_x(x,t). \end{aligned} \tag{9}$$

The system (6) and (7) rewritten in terms of w and y is given by

$$\begin{aligned} w_{xt} + F[w,y] = 0, \quad & F[w,y] = w_x w_{xx} + w_{xxxx} + \lambda y_x y_{xx} \\ y_{xt} + G[w,y] = 0, \quad & G[w,y] = w_{xx} y_x + y_{xx} w_x + y_{xxxx}. \end{aligned} \tag{10}$$

We notice that the matrix constructed from the Frechet derivatives of F and G, with respect to w and y, is self-adjoint. We then conclude from the Helmholtz procedure that

$$\mathcal{L}_1 = -\frac{1}{2}w_x w_t - \frac{1}{2}\lambda y_x y_t + \int_0^1 (wF[\mu w, \mu y] + y\lambda G[\mu w, \mu y])d\mu, \tag{11}$$

where $\lambda \neq 0$, and

$$\mathcal{L}_2 = -\frac{1}{2}w_x y_t - \frac{1}{2}w_t y_x + \int_0^1 (yF[\mu w, \mu y] + wG[\mu w, \mu y])d\mu, \tag{12}$$

for every real value of λ, are two Lagrangian densities which give rise, from a variational principle to Eqs. (6) and (7).

The Lagrangians associated with $\mathcal{L}_i$, $i = 1, 2$ are given by $L_i(w, y) = \int_0^T dt \int_{-\infty}^{+\infty} dx \mathcal{L}_i$, $i = 1, 2$.

Independent variations of L_i, for each i, with respect to w and y give rise to the field equations

$$\begin{aligned} \delta_w L_i &= 0 \\ \delta_y L_i &= 0 \end{aligned} \tag{13}$$

which coincide, for each i, with Eqs. (6) and (7). In the above equations δ_w and δ_y denote the Gateaux functional variation defined by

$$\begin{aligned} \delta_w L(w, y) &= \lim_{\epsilon \to 0} \frac{L(w + \epsilon\delta w, y) - L(w, y)}{\epsilon} \\ \delta_y L(w, y) &= \lim_{\epsilon \to 0} \frac{L(w, y + \epsilon\delta y) - L(w, y)}{\epsilon}. \end{aligned} \tag{14}$$

The explicit expressions for $\mathcal{L}_1$ and $\mathcal{L}_2$ are given by

$$\mathcal{L}_1 = -\frac{1}{2}w_x w_t - \frac{1}{6}{w_x}^3 + \frac{1}{2}{w_{xx}}^2 - \frac{\lambda}{2}w_x {y_x}^2 - \frac{\lambda}{2}y_x y_t + \frac{\lambda}{2}{y_{xx}}^2, \tag{15}$$

$$\mathcal{L}_2 = -\frac{1}{2}w_x y_t - \frac{1}{2}w_t y_x - \frac{1}{2}w_x^2 y_x - y_x w_{xxx} - \frac{\lambda}{6}y_x^3. \tag{16}$$

The Lagrangians L_i, $i = 1, 2$, are singular Lagrangians, we thus expect a constrained Hamiltonian formulation associated with them. The same happens for the corresponding KdV Lagrangian that can be obtained from $\mathcal{L}_1$ by imposing $\lambda = 0$.

We consider first the Lagrangian L_1. The conjugate momenta associated with w and y, which we denote by p and q, respectively, are given by

$$p = \frac{\partial \mathcal{L}_1}{\partial w_t} = -\frac{1}{2}w_x, \quad q = \frac{\partial \mathcal{L}_1}{\partial y_t} = -\frac{\lambda}{2}y_x. \tag{17}$$

We define

$$\phi_1 \equiv p + \frac{1}{2}w_x, \quad \phi_2 = q + \frac{\lambda}{2}y_x. \tag{18}$$

Hence, $\phi_1 = \phi_2 = 0$ are constraints on the phase space. We then follow the Dirac procedure to determine the whole set of constraints. It turns out that these are the only constraints on the phase space.

The Hamiltonian density may be obtained directly from $\mathcal{L}_1$ by performing a Legendre transformation,

$$\mathcal{H}_1 = pw_t + qy_t - \mathcal{L}_1. \tag{19}$$

The Hamiltonian density is then given by

$$\mathcal{H}_1 = \frac{1}{6}w_x^3 - \frac{1}{2}w_{xx}^2 + \frac{\lambda}{2}w_x y_x^2 - \frac{\lambda}{2}y_{xx}^2 \tag{20}$$

and the Hamiltonian by $H_1 = \int_{-\infty}^{+\infty} dx\, \mathcal{H}_1$.

We introduce a Poisson structure on the phase space by defining

$$\begin{aligned} \{w(x), p(\hat{x})\}_{PB} &= \delta(x-\hat{x}) \\ \{y(x), q(\hat{x})\}_{PB} &= \delta(x-\hat{x}) \end{aligned} \tag{21}$$

with all other brackets between these variables being zero.

From them we obtain

$$\{\partial_x^n w(x), \partial_{\hat{x}}^m p(\hat{x})\} = \partial_x^n \partial_{\hat{x}}^m \{w(x), p(\hat{x})\}. \tag{22}$$

It turns out that ϕ_1, ϕ_2 are second class constraints. In fact,

$$\begin{aligned} \{\phi_1(x), \phi_1(\hat{x})\}_{PB} &= \delta_x(x-\hat{x}) \\ \{\phi_1(x), \phi_2(\hat{x})\}_{PB} &= 0 \\ \{\phi_2(x), \phi_2(\hat{x})\}_{PB} &= \lambda\delta_x(x-\hat{x}). \end{aligned} \tag{23}$$

In order to define the Poisson structure on the constrained phase space, we need to use the Dirac brackets.

The Dirac bracket between two functionals F and G on phase space is defined by

$$\{F, G\}_{DB} = \{F, G\}_{PB} - \langle\langle \{F, \phi_i(x')\}_{PB} \mathbb{C}_{ij}(x', x'') \{\phi_j(x''), G\}_{PB} \rangle_{x'} \rangle_{x''} \tag{24}$$

where $<>_{x'}$ denotes integration on x' from $-\infty$ to $+\infty$. The indices $i, j = 1, 2$ and the $\mathbb{C}_{ij}(x', x'')$ are the components of the inverse of the matrix whose components are $\{\phi_i(x'), \phi_j(x''\}_{PB}$.

This matrix becomes

$$\begin{bmatrix} \partial_{x'}\delta(x'-x'') & 0 \\ 0 & \lambda\partial_{x'}\delta(x'-x'') \end{bmatrix} \tag{25}$$

and its inverse is given by

$$[\mathbb{C}_{ij}(x', x'')] = \begin{bmatrix} \int^{x'} \delta(s-x'')ds & 0 \\ 0 & \frac{1}{\lambda}\int^{x'} \delta(s-x'')ds \end{bmatrix}. \tag{26}$$

It turns out, after some calculations, that the Dirac brackets of the original variables are

$$\begin{aligned} &\{u(x), u(\hat{x})\}_{DB} = -\partial_x \delta(x-\hat{x}), \quad \{v(x), v(\hat{x})\}_{DB} = -\frac{1}{\lambda}\partial_x \delta(x-\hat{x}) \\ &\{u(x), v(\hat{x})\}_{DB} = 0. \end{aligned} \tag{27}$$

We remind that this Poisson structure has been constructed assuming $\lambda \neq 0$.

From them, we obtain the Hamilton equations, which of course are the same as Eqs. (6) and (7):

$$\begin{aligned} u_t &= \{u, H_1\}_{DB} = -uu_x - u_{xxx} - \lambda v v_x \\ v_t &= \{v, H_1\}_{DB} = -u_x v - v_x u - v_{xxx}. \end{aligned} \tag{28}$$

We notice that adding any function of the constraints to H_1 does not change the result, since the Dirac bracket of the constraints with any other local function of the phase space variables is zero.

Using the above bracket relations for u and v, we may obtain directly the Dirac bracket of any two functionals $F(u, v)$ and $G(u, v)$. We notice that the observables F and G may be functionals of w, y, p, and q, not only of u and v. In this sense, the phase space approach for singular Lagrangians provides the most general space of observables.

We now consider the action L_2 and its associated Hamiltonian structure. In this case, we denote the conjugate momenta to w and y by $\hat{p}$ and $\hat{q}$, respectively. We have

$$\hat{p} = -\frac{1}{2}y_x, \quad \hat{q} = -\frac{1}{2}w_x. \tag{29}$$

In this case, the constraints become

$$\widehat{\phi_1} = \hat{p} + \frac{1}{2}y_x = 0, \quad \widehat{\phi_2} = \hat{q} + \frac{1}{2}w_x = 0. \tag{30}$$

The corresponding Poisson brackets are given by

$$\begin{aligned} &\{\widehat{\phi_1}(x), \widehat{\phi_1}(x')\}_{PB} = 0, \quad \{\widehat{\phi_2}(x), \widehat{\phi_2}(x')\}_{PB} = 0, \\ &\{\widehat{\phi_1}(x), \widehat{\phi_2}(x')\}_{PB} = \partial_x \delta(x-x'). \end{aligned} \tag{31}$$

From them, we can construct the Dirac brackets after which some calculations yield the Poisson structure for the original variables

http://dx.doi.org/10.5772/66146

$$\begin{aligned}&\{u(x),u(\hat{x})\}_{DB}=0,\quad \{v(x),v(\hat{x})\}_{DB}=0,\\&\{u(x),v(\hat{x})\}_{DB}=-\partial_x\delta(x-\hat{x}).\end{aligned} \tag{32}$$

The Hamiltonian $H_2=\int_{-\infty}^{+\infty}dx\,\mathcal{H}_2$ is given in terms of the Hamiltonian density

$$\mathcal{H}_2=\frac{1}{2}w_x^2y_x+y_xw_{xxx}+\frac{\lambda}{6}y_x^3. \tag{33}$$

The Hamilton equations follow then in terms of the Dirac brackets, they are

$$u_t=\{u,H_2\}_{DB},\quad v_t=\{v,H_2\}_{DB}, \tag{34}$$

which coincide with the field Eqs. (6) and (7) for any value of λ. We have thus constructed two Hamiltonian functionals and associated Poisson bracket structures. These two Hamiltonian structures arise directly from the basic actions L_1 and L_2. In Section 6, we will construct two additional Hamiltonian structures by considering a Miura transformation for the coupled system.

5. A pencil of Poisson structures for the parametric coupled KdV system

We have then constructed two Lagrangian densities $\mathcal{L}_i$, $i=1,2$, we may now introduce a real parameter k and define a parametric Lagrangian density

$$\mathcal{L}_k=k\mathcal{L}_1+(1-k)\mathcal{L}_2. \tag{35}$$

The field equations obtained from the corresponding Lagrangian $L_k=\int_0^T dt\int_{-\infty}^{+\infty}dx\mathcal{L}_k$ are equivalent to Eqs. (6) and (7) in the following cases: If $\lambda<0$ for any k. If $\lambda=0$, for $k\neq1$. If $\lambda>0$ for $k\neq\frac{1}{1+\sqrt{\lambda}}$ and $k\neq\frac{1}{1-\sqrt{\lambda}}$. From now on, we will exclude these particular values of k.

The parametric Lagrangian L_k is singular for any value of k (excluding the above mentioned particular cases). The corresponding phase space formulation contains constraints, which are determined by the use of the Dirac procedure. We denote p and q the conjugate momenta associated with w and y, respectively. From their definition, we obtain the primary constraints.

$$\phi_1\equiv\frac{k}{2}w_x+\frac{(1-k)}{2}y_x+p=0 \tag{36}$$

$$\phi_2\equiv\frac{\lambda k}{2}y_x+\frac{(1-k)}{2}w_x+q=0. \tag{37}$$

We may then define the Hamiltonian density $\mathcal{H}_k$ through the Legendre transformation, we get

$$\mathcal{H}_k = pw_t + qy_t - \mathcal{L}_k = k\mathcal{H}_1 + (1-k)\mathcal{H}_2. \tag{38}$$

We now follow the Dirac algorithm to determine the complete set of constraints. It turns out that these are the only constraints in the formulation.

The Poisson brackets of the constraints obtained from the canonical Poisson brackets of the conjugate pairs are

$$\begin{aligned}\{\phi_1(x),\phi_1(\hat{x})\}_{PB} &= k\partial_x\delta(x-\hat{x})\\ \{\phi_2(x),\phi_2(\hat{x})\}_{PB} &= \lambda k\partial_x\delta(x-\hat{x})\\ \{\phi_1(x),\phi_2(\hat{x})\}_{PB} &= (1-k)\partial_x\delta(x-\hat{x}).\end{aligned} \tag{39}$$

Hence, they are second class constraints. We will denote by $\{\}^k_{DB}$ the Dirac bracket corresponding to the parameter k. We then proceed to calculate the Dirac brackets of the original fields u and v.

We obtain

$$\begin{aligned}\{u(x),u(\hat{x})\}^k_{DB} &= \frac{\lambda k}{-\lambda k^2 + (1-k)^2}\partial_x\delta(x-\hat{x})\\ \{v(x),v(\hat{x})\}^k_{DB} &= \frac{k}{-\lambda k^2 + (1-k)^2}\partial_x\delta(x-\hat{x})\\ \{u(x),v(\hat{x})\}^k_{DB} &= \frac{1-k}{-\lambda k^2 + (1-k)^2}\left(-\partial_x\delta(x-\hat{x})\right).\end{aligned} \tag{40}$$

where the denominator is different from zero for the values of k we are considering. The above brackets define the Poisson structure of the corresponding Hamiltonian

$$H_k = \int_{-\infty}^{+\infty} dx\mathcal{H}_k. \tag{41}$$

The Hamilton equations

$$\begin{aligned}u_t &= \{u, H_k\}_{DB}\\ v_t &= \{v, H_k\}_{DB}\end{aligned} \tag{42}$$

coincide, as expected, with the coupled Eqs. (6) and (7).

In Section 3, we constructed two Poisson structures for the coupled system (6) and (7). We now show they are compatible. It follows, for any two functionals F and G that

$$\{F, G\}^k_{DB} = \frac{-\lambda k}{-\lambda k^2 + (1-k)^2}\{F, G\}^1_{DB} + \frac{1-k}{-\lambda k^2 + (1-k)^2}\{F, G\}^0_{DB}, \tag{43}$$

where $\{F, G\}^1_{DB}$, corresponding to $k = 1$, and $\{F, G\}^0_{DB}$, corresponding to $k = 0$, are the two Dirac brackets structures obtained in Section 3. In particular, for any $\lambda \neq 0, 1$ and $k = \frac{1}{1-\lambda}$, we get

$$\{F,G\}_{DB}^{k} = \{F,G\}_{DB}^{1} + \{F,G\}_{DB}^{0}, \tag{44}$$

which implies that any linear combination of $\{F,G\}_{DB}^{1}$ and $\{F,G\}_{DB}^{0}$, for any $\lambda \neq 0, 1$, is a Poisson bracket. That is, the two Poisson structures obtained in Ref. [22], corresponding to $k = 1$ and $k = 0$, are compatible.

For the particular value of $\lambda = 0$, and any $k \neq 1$ we obtain

$$\{F,G\}_{DB}^{k} = \frac{k}{2(1-k)^2}\{F,G\}_{DB}^{\frac{1}{2}} + \frac{1-2k}{(1-k)^2}\{F,G\}_{DB}^{0}. \tag{45}$$

For $k = \frac{2}{5}$ the two coefficients on the right-hand member of Eq. (45) are equal. It implies that the Poisson structures for $k = \frac{1}{2}$ and $k = 0$ are compatible.

We have thus constructed a pencil of Poisson structures, except for $\lambda = 1$, for which the coupled system reduces to two decoupled KdV equations.

6. The Miura transformation for the parametric coupled KdV system

It is well known that the KdV equation admits two Hamiltonian structures, one of them is a particular case of our previous construction. It is obtained by considering only the $u(x, t)$ field, imposing $v(x, t) = 0$. In this case, the two previous Hamiltonians structures reduce to only one and there is no pencil of Poisson structures. The second Hamiltonian structure for the KdV equation arises from a Miura transformation, which is also a particular case of the following construction. The corresponding Miura transformation for our coupled system becomes

$$\begin{aligned} u &= \mu_x - \frac{1}{6}\mu^2 - \frac{\lambda}{6}\nu^2 \\ v &= \nu_x - \frac{1}{3}\mu\nu. \end{aligned} \tag{46}$$

and the modified KdV system (MKdVS)

$$\begin{aligned} &\mu_t + \mu_{xxx} - \frac{1}{6}\mu^2\mu_x - \frac{\lambda}{6}\nu^2\mu_x - \frac{\lambda}{3}\mu\nu\nu_x = 0 \\ &\nu_t + \nu_{xxx} - \frac{1}{6}\mu^2\nu_x - \frac{\lambda}{6}\nu^2\nu_x - \frac{1}{3}\mu\nu\mu_x = 0. \end{aligned} \tag{47}$$

It is interesting that from Eq. (47), following the Helmholtz procedure, which is also valid for the MKdVS system, we obtain two singular Lagrangians densities $\mathcal{L}_i^M, i = 1,2,$ expressed in terms of the Casimir potentials σ, ρ where $\mu = \sigma_x, \nu = \rho_x$:

$$\mathcal{L}_1^M = -\frac{1}{2}\sigma_t\sigma_x - \frac{\lambda}{2}\rho_t\rho_x - \frac{1}{2}\sigma_x\sigma_{xxx} - \frac{\lambda}{2}\rho_x\rho_{xxx} + \frac{1}{72}{\sigma_x}^4 + \frac{\lambda^2}{72}{\rho_x}^4 + \frac{\lambda}{12}\rho_x^2\sigma_x^2 \tag{48}$$

and

$$\mathcal{L}_2^M = -\frac{1}{2}\sigma_t\rho_x - \frac{1}{2}\sigma_x\rho_t - \sigma_{xxx}\rho_x + \frac{1}{18}{\sigma_x}^3\rho_x + \frac{\lambda}{18}{\rho_x}^3\sigma_x, \tag{49}$$

Eq. (48) being valid only for $\lambda \neq 0$.

Each of them has a Poisson structure that follows from the Dirac approach. The Dirac brackets, for the original fields u, v in the coupled system (6) and (7) are given by

$$\begin{aligned} \{u(x),u(\hat{x})\}_{DB} &= \partial_{xxx}\delta(x-\hat{x}) + \frac{1}{3}u_x\delta(x-\hat{x}) + \frac{2}{3}u\partial_x\delta(x-\hat{x}) \\ \{v(x),v(\hat{x})\}_{DB} &= \frac{1}{\lambda}\partial_{xxx}\delta(x-\hat{x}) + \frac{1}{3\lambda}u_x\delta(x-\hat{x}) + \frac{2}{3\lambda}u\partial_x\delta(x-\hat{x}) \\ \{u(x),v(\hat{x})\}_{DB} &= \frac{1}{3}v_x\delta(x-\hat{x}) + \frac{2}{3}v\partial_x\delta(x-\hat{x}), \end{aligned} \tag{50}$$

which is the Poisson structure associated with $\mathcal{L}_1^M$ and

$$\begin{aligned} \{u(x),u(\hat{x})\}_{DB} &= \frac{\lambda}{3}v_x\delta(x-\hat{x}) + \frac{2\lambda}{3}v\partial_x\delta(x-\hat{x}) \\ \{v(x),v(\hat{x})\}_{DB} &= \frac{1}{3}v_x\delta(x-\hat{x}) + \frac{2}{3}v\partial_x\delta(x-\hat{x}) \\ \{u(x),v(\hat{x})\}_{DB} &= \partial_{xxx}\delta(x-\hat{x}) + \frac{1}{3}u_x\delta(x-\hat{x}) + \frac{2}{3}u\partial_x\delta(x-\hat{x}), \end{aligned} \tag{51}$$

the Poisson structure associated with $\mathcal{L}_2^M$.

The corresponding Hamiltonian densities $\mathcal{H}_1^M$ and $\mathcal{H}_2^M$ are given in terms of the fields u and v by

$$\begin{aligned} \mathcal{H}_1^M &= v^2 - u^2 \\ \mathcal{H}_1^M &= -uv. \end{aligned} \tag{52}$$

The Hamilton equations obtained from these Hamiltonian structures coincide, of course, with Eqs. (6) and (7).

From these two Poisson structures, we may construct a pencil of Poisson structures as we described in the previous section, see Ref. [22] for the details of the construction. We notice that $\mathcal{L}_1^M$ and $\mathcal{L}_2^M$ in the construction are of the same dimension. It is then not possible to construct a hierarchy of higher order Hamiltonians from them. The same occurs with $\mathcal{L}_1$ and $\mathcal{L}_2$. However, the two pencils are of different dimensions and we may obtain from them a hierarchy of higher order Hamiltonians which extends the hierarchy of the KdV equation.

7. A duality relation among the Lagrangians of the parametric coupled KdV system

We consider a generalization of the Gardner construction for the KdV equation. The Gardner transformation for the system (6) and (7) is given by

$$u = r + \varepsilon r_x - \frac{1}{6}\varepsilon^2 (r^2 + \lambda s^2) \tag{53}$$

$$v = s + \varepsilon s_x - \frac{1}{3}\varepsilon^2 rs, \tag{54}$$

where ε is a real parameter and $r(x, t), s(x, t)$ are the fields which describe the Gardner ε-deformation. The Gardner equations are

$$r_t + r_{xxx} + r r_x + \lambda s s_x - \frac{1}{6}\varepsilon^2 [(r^2 + \lambda s^2) r_x + 2\lambda r s s_x] = 0 \tag{55}$$

$$s_t + s_{xxx} + r s_x + s r_x - \frac{1}{6}\varepsilon^2 [(r^2 + \lambda s^2) s_x + 2 r s r_x] = 0. \tag{56}$$

Any solution of Eqs. (55) and (56) define through Eqs. (53) and (54) a solution of the system (6), (7).

$\int_{-\infty}^{+\infty} dx\, r(x, t)$ and $\int_{-\infty}^{+\infty} dx\, s(x, t)$ are conserved quantities of the system (55) and (56). Assuming a formal power series on ε of the solutions of Eqs. (55) and (56) and inverting Eqs. (53) and (54), one obtains an infinite sequence of conserved quantities for the system (6), (7). It is an integrable system in this sense.

If we consider the $\varepsilon \to 0$ limit for the Gardner transformation Eqs. (53), (54) and Gardner Eqs. (55) and (56), we get the original system (6) and (7). On the other side, if we redefine

$$\mu \equiv \varepsilon r \tag{57}$$

$$\nu \equiv \varepsilon s \tag{58}$$

and rewrite Eqs. (53) and (54), we get

$$u = \frac{\mu}{\varepsilon} + \mu_x - \frac{1}{6}\mu^2 - \frac{1}{6}\lambda \nu^2 \tag{59}$$

$$v = \frac{\nu}{\varepsilon} + \nu_x - \frac{1}{3}\mu\nu. \tag{60}$$

Taking the limit $\varepsilon \to \infty$ we obtain

$$\hat{u} = \mu_x - \frac{1}{6}\mu^2 - \frac{1}{6}\lambda\mu^2 \tag{61}$$

$$\hat{v} = \nu_x - \frac{1}{3}\mu\nu \tag{62}$$

which is exactly the Miura transformation. In the same limit, we obtain from Eqs. (55), (56) the Miura equations given by Eq. (47).

We now construct using the Helmholtz approach a master Lagrangian for the Gardner equations. The master Lagrangians, there are two of them, are ε dependent and following the above limits we obtain all the Lagrangian structures we discussed previously. The KdV coupled system and the modified KdV coupled system are then dual constructions corresponding to the weak coupling limit $\varepsilon \to 0$ and to the strong coupling limit $\varepsilon \to \infty$ respectively, of the master construction. A direct relation of these two systems arises from the present construction.

We introduce the Casimir potentials

$$r = w_x, \quad s = y_x \tag{63}$$

and using the Helmholtz approach we obtain the Lagrangian densities

$$\begin{aligned} \mathcal{L}_{G1} = &-\frac{1}{2}w_x w_t - \frac{1}{6}(w_x)^3 + \frac{1}{2}(w_{xx})^2 - \frac{\lambda}{2}w_x(y_x)^2 - \frac{\lambda}{2}y_x y_t + \frac{\lambda}{2}(y_{xx})^2 \\ &-\frac{1}{6}\varepsilon^2\left[-\frac{1}{12}(w_x)^4 - \frac{\lambda}{2}(w_x)^2(y_x)^2\right] + \frac{\varepsilon^2}{72}\lambda^2(y_x)^4, \end{aligned} \tag{64}$$

$$\begin{aligned} \mathcal{L}_{G2} = &-\frac{1}{2}w_x y_t - \frac{1}{2}w_t y_x - \frac{1}{2}(w_x)^2 y_x - y_x w_{xxx} - \frac{\lambda}{6}(y_x)^3 \\ &+\frac{1}{18}\varepsilon^2(w_x)^3 y_x + \frac{1}{18}\varepsilon^2\lambda(y_x)^3 w_x. \end{aligned} \tag{65}$$

If we take the weak coupling limit $\varepsilon \to 0$ we obtain

$$\lim_{\varepsilon\to 0}\mathcal{L}_{G1} = \mathcal{L}_1 \quad , \quad \lim_{\varepsilon\to 0}\mathcal{L}_{G1} = \mathcal{L}_2 \tag{66}$$

where $\mathcal{L}_1$ and $\mathcal{L}_2$ were defined in Section 3.

If we redefine

$$\begin{aligned} \sigma = \varepsilon w \quad &, \quad \rho = \varepsilon y \\ \mathcal{L}^M_{G1} = \varepsilon^2\mathcal{L}_{G1} \quad &, \quad \mathcal{L}^M_{G2} = \varepsilon^2\mathcal{L}_{G2} \end{aligned} \tag{67}$$

and take the strong coupling limit $\varepsilon \to \infty$, we get

$$\begin{aligned} \lim_{\varepsilon\to\infty}\mathcal{L}^M_{G1}(\sigma,\rho) = \mathcal{L}^M_1(\sigma,\rho) \\ \lim_{\varepsilon\to\infty}\mathcal{L}^M_{G2}(\sigma,\rho) = \mathcal{L}^M_2(\sigma,\rho), \end{aligned} \tag{68}$$

where $\mathcal{L}_1^M$ and $\mathcal{L}_2^M$ were defined in Section 5. Consequently, all the Lagrangian structure and the associated Hamiltonian structure of the coupled system (6), (7) arises from the master Lagrangians. They can also be combined to a unique master Lagrangian depending on a parameter k as was done in Section 4. The field equations of the master Lagrangians are the Gardner equations, the spatial integral of $r(x, t)$ and $s(x, t)$ define an ε-deformed conserved quantity of the Gardner equations which implies an infinite sequence of conserved quantities of the original coupled KdV system (6), (7).

8. Hamiltonian structure for a KdV system valued on a Clifford algebra

In this section, we continue the discussion of the Lagrangian and Hamiltonian structures for the coupled KdV systems. We discuss a coupled system arising from the breaking of the supersymmetry on the $N = 1$ supersymmetric KdV equation. The details of this system may be found in Ref. [15]. The system is formulated in terms of a real valued field $u(x, t)$ and a Clifford algebra valued field $\xi(x, t)$. The field $\xi(x, t)$ is expressed in terms of an odd number of generators e_i, $i = 1, \ldots$ of the Clifford algebra

$$\xi = \sum_{i=1}^{\infty} \varphi_i e_i + \sum_{ijk} \varphi_{ijk} e_i e_j e_k + \cdots \tag{69}$$

where

$$e_i e_j + e_j e_i = -2\delta_{ij}, \tag{70}$$

and φ_i, φ_{ijk}, … are real valued fields. We define by $\overline{\xi}$ the conjugate of ξ,

$$\overline{\xi} = \sum_{i=1}^{\infty} \varphi_i \overline{e}_i + \sum_{ijk} \varphi_{ijk} \overline{e}_k \overline{e}_j \overline{e}_i + \cdots \tag{71}$$

where $\overline{e}_i = -e_i$. We denote by $\mathcal{P}(\xi\overline{\xi})$ the projector of the product $\xi\overline{\xi}$ to the identity element of the algebra

$$\mathcal{P}(\xi\overline{\xi}) = \sum_{i=1}^{\infty} \varphi_i^2 + \sum_{ijk} \varphi_{ijk}^2 + \cdots \tag{72}$$

We proposed in Ref. [15] the following coupled KdV system arising from the breaking of the supersymmetry in the $N = 1$ supersymmetric equation [9]:

$$\begin{aligned} u_t &= -u_{xxx} - u u_x - \frac{1}{4}\left(\mathcal{P}(\xi\overline{\xi})\right)_x \\ \xi_t &= -\xi_{xxx} - \frac{1}{2}(\xi u)_x. \end{aligned} \tag{73}$$

In distinction to the $N=1$ supersymmetric KdV equation the coupled system (73) has only a finite number of local conserved quantities,

$$\begin{aligned}
\hat{H}_{\frac{1}{2}} &= \int_{-\infty}^{+\infty} \xi dx, \\
\hat{H}_1 &= \int_{-\infty}^{+\infty} u dx, \\
V \equiv \hat{H}_3 &= \int_{-\infty}^{+\infty} \left(u^2 + \mathcal{P}(\xi\overline{\xi})\right) dx, \\
M \equiv \hat{H}_5 &= \int_{-\infty}^{+\infty} \left(-\frac{1}{3}u^3 - \frac{1}{2}u\mathcal{P}(\xi\overline{\xi}) + (u_x)^2 + \mathcal{P}(\xi_x\overline{\xi}_x)\right) dx.
\end{aligned} \tag{74}$$

It is interesting to remark that the following nonlocal conserved charge of Super KdV [32] is also a nonlocal conserved charge for the system (73), in terms of the Clifford algebra valued field ξ,

$$\int_{-\infty}^{\infty} \xi(x) \int_{-\infty}^{x} \xi(s) ds dx. \tag{75}$$

However, the nonlocal conserved charges of Super KdV in Ref. [33] are not conserved by the system (73). For example,

$$\int_{-\infty}^{\infty} u(x) \int_{-\infty}^{x} \xi(s) ds dx. \tag{76}$$

is not conserved by Eq. (73).

The system (73) has multisolitonic solutions. In Ref. [34], we showed that the soliton solution is Liapunov stable under perturbation of the initial data.

9. The Lagrangian and Hamiltonian structure of the Clifford valued system

We introduce the Casimir potentials w and η defined by

$$u = w_x \text{ and } \xi = \eta_x. \tag{77}$$

We notice, as in the previous sections, that Eq. (73) may be expressed as stationary points of a singular Lagrangian constructed following the Helmholtz approach. We denote

$$\begin{aligned}
P(w,\eta) &= w_{xxxx} + w_x w_{xx} + \frac{1}{4}\left(\mathcal{P}(\eta_x\overline{\eta}_x)\right)_x \\
Q(w,\eta) &= \eta_{xxxx} + \frac{1}{2}(w_x\eta_x)_x
\end{aligned} \tag{78}$$

The Lagrangian becomes $L = \int_0^T dt \int_{-\infty}^{+\infty} dx \mathcal{L}$ in terms of the Lagrangian density $\mathcal{L}$ given by

$$\mathcal{L} = \frac{1}{2}w_x w_t + \frac{1}{2}(\mathcal{P}(\eta_x\overline{\eta}_t)) - \int_0^1 wP(\mu w, \mu\eta)d\mu - \int_0^1 \mathcal{P}(Q(\mu w, \mu\eta)\overline{\eta})d\mu. \quad (79)$$

From the Lagrangian L, we may construct its Hamiltonian structure using the Legendre transformation. We denote (p, σ) the conjugate momenta to (w, η):

$$\begin{aligned} p &:= \frac{\partial\mathcal{L}}{\partial(\partial_t w)} = \frac{1}{2}w_x = \frac{1}{2}u \\ \sigma &:= \frac{\partial\mathcal{L}}{\partial(\partial_t \eta)} = \frac{1}{2}\eta_x = \frac{1}{2}\varphi. \end{aligned} \quad (80)$$

Eq. (80) describes constraints on the phase space.

Performing the Legendre transformation we obtain the Hamiltonian of the system

$$H = \int_{-\infty}^{+\infty} dx\left(pw_t + \mathcal{P}(\sigma\overline{\eta}_t) - \mathcal{L}\right) \quad (81)$$

where $H = \frac{1}{2}\hat{H}_5$ in (74).

Following the Dirac approach, the conservation of the primary constraints (80) determines the Lagrange multipliers associated with the constraints (80). There are no more constraints on the phase space. It turns out that both constraints are second class ones. The Poisson structure of the constrained Hamiltonian is then determined by the Dirac brackets, see Ref. [15] for the details. We identify by an index i the independent components of a field η or σ valued on the Clifford algebra. We may rewrite the constraints as

$$\begin{aligned} v &:= p - \frac{1}{2}w_x \\ v_i &:= \sigma_i - \frac{1}{2}\eta_{ix}. \end{aligned} \quad (82)$$

Introducing $v_I := (v, v_i)$, we then have

$$\{v_I(x), v_J(x')\} = -\delta_{IJ}\partial_x\delta(x - x'). \quad (83)$$

The Poisson structure of the constrained Hamiltonian is then determined by the Dirac brackets [20]. For any two functionals on the phase space F and G, the Dirac bracket is defined as

$$\{F, G\}_{DB} := \{F, G\} - \langle\langle\{F, v_I(x')\}\{v_I(x'), v_J(x'')\}^{-1}\rangle_{x'}\{v_J(x''), G\}\rangle_{x''}, \quad (84)$$

where

$$\langle\{v_I(x'), v_J(x'')\}^{-1}g(x'')\rangle_{x''} = -\delta_{IJ}\int_{-\infty}^{x'} g(\tilde{x})d\tilde{x}. \quad (85)$$

We then have

$$\begin{aligned}&\{u(x),u(y)\}_{DB}=\partial_x\delta(x,y),\\&\{\varphi_i(x),\varphi_j(y)\}_{DB}=\delta_{ij}\partial_x\delta(x,y),\\&\{u(x),\varphi_i(y)\}_{DB}=0.\end{aligned} \tag{86}$$

Consequently,

$$\begin{aligned}\partial_t u&=\{u,H\}_{DB}=-\frac{1}{2}(u^2)_x-u_{xxx}-\frac{\lambda}{4}(\varphi_i^2)_x\\ \partial_t\varphi_i&=\{\varphi_i,H\}_{DB}=-\varphi_{ixxx}-\frac{\lambda}{2}(u\varphi_i)_x,\end{aligned} \tag{87}$$

where H is given by the last conserved quantity in Eq. (74) and can be directly expressed in terms of u and ξ.

10. Positiveness of the Hamiltonian for the Clifford valued system

An interesting property of the Hamiltonian H of the Clifford coupled system (73) is its *a priori* positiveness. In fact,

$$\hat{H}_3+\hat{H}_5=\|(u,\xi)\|_{H_1}^2+\int_{-\infty}^{+\infty}\left(-\frac{1}{3}u^3-\frac{1}{2}u\mathcal{P}(\xi\overline{\xi})\right)dx \tag{88}$$

where the Sobolev norm $\|\|_{H_1}$ is defined by

$$\|(u,\xi)\|_{H_1}^2:=\int_{-\infty}^{+\infty}[u^2+\mathcal{P}(\xi\overline{\xi})+u_x{}^2+\mathcal{P}(\xi_x\overline{\xi}_x)]dx. \tag{89}$$

We also noticed that

$$\hat{H}_3=\|(u,\xi)\|_{L^2}^2 \tag{90}$$

where $\|\|_{L^2}$ is the L^2 norm.

We then have

$$\hat{H}_3+\hat{H}_5\geq\|(u,\xi)\|_{H_1}^2-\frac{1}{2}\int_{-\infty}^{+\infty}|u|\left(u^2+\mathcal{P}(\xi\overline{\xi})\right)dx. \tag{91}$$

We now use the bound

$$\sup|u|\leq\frac{\|u\|_{H_1}}{\sqrt{2}}\leq\frac{\|(u,\xi)\|_{H_1}}{\sqrt{2}}, \tag{92}$$

to obtain

$$\hat{H}_3+\hat{H}_5\geq\|(u,\xi)\|_{H_1}^2-\frac{1}{2\sqrt{2}}\|(u,\xi)\|_{H_1}\|(u,\xi)\|_{L^2}. \tag{93}$$

Consequently,

$$\hat{H}_3 + \hat{H}_5 + \left(\frac{1}{4\sqrt{2}}\right)^2 \hat{H}_3 \geq \left(\|(u,\xi)\|_{H_1} - \frac{1}{4\sqrt{2}}\|(u,\xi)\|_{L^2}\right)^2 \geq 0. \tag{94}$$

Finally,

$$\hat{H}_5 \geq -\left(1 + \left(\frac{1}{4\sqrt{2}}\right)^2\right)\hat{H}_3, \tag{95}$$

Hence, for a normalized state satisfying $\|(u,\xi)\|_{L^2} = 1$, we have

$$\hat{H}_5 \geq -\left(1 + \left(\frac{1}{4\sqrt{2}}\right)^2\right). \tag{96}$$

The Hamiltonian is then manifestly bounded from below in the space of normalized L_2 configurations and it is thus physically admissible.

The property of the Hamiltonian is relevant from the physical point of view. In particular, in showing that the soliton solution of the Clifford coupled system is Liapunov stable. The stability analysis follows ideas introduced in Ref. [35] for the KdV equation. It is based on the use of the conserved quantities of the system. It is interesting that only the first few of them, in the case of the KdV equation, are needed. In the case of the Clifford coupled system these are all the local conserved quantities of the system.

11. The KdV equation valued on the octonion algebra

A famous theorem by Hurwitz establishes that the only real normalized division algebras are the reals $\mathbb{R}$, the complex $\mathbb{C}$, the quaternions $\mathbb{H}$, and the octonions O. In particular, these division algebras are directly related to the existence of super Yang-Mills in several dimensions: 3, 4, 6, and 10 dimensions [36]. The octonion algebra may be explicitly used in the formulation of superstring theory in 10 dimensions and in the supermembrane theory in 11 dimensions, relevant theories in the search for a unified theory of all the known fundamental forces in nature.

The extension of the KdV equation to a partial differential equation for a field valued on a octonion algebra is then an interesting goal [23].

We showed in the previous sections that an extension of the KdV equation to the field valued on a Clifford algebra give rise to a coupled system with Liapunov stable soliton solution but without an infinite sequence of local conserved quantities.

In the present section, we analyze the KdV extension where the field is valued on the octonion algebra. The system shares several properties of the original real KdV equation. It has soliton solutions and also has an infinite sequence of local conserved quantities derived from a

Bäcklund transformation and a bi-Lagrangian and bi-Hamiltonian structure [23]. We will show in this section the construction of the bi-Lagrangian structure.

The octonion algebra contains as subalgebras all other division algebras, hence our construction may be reduced to any of them.

The KdV equation on the octonion algebra can be seen as a coupled KdV system, as we will see it has some similarities to the construction in the previous sections. However, it is invariant under the exceptional Lie group G_2, the automorphisms of the octonions, and under the Galileo transformations. Those symmetries are not present in the model constructed on a Clifford algebra.

We denote $u = u(x, t)$ a function with domain in $\mathbb{R}\times\mathbb{R}$ valued on the octonionic algebra. If we denote e_i, $i = 1, \ldots, 7$ the imaginary basis of the octonions, u can be expressed as

$$u(x, t) = b(x, t) + \vec{B}(x, t) \tag{97}$$

where $b(x, t)$ is the real part and $\vec{B} = \sum_{i=1}^{7} B_i(x, t) e_i$ its imaginary part.

The KdV equation formulated on the algebra of octonions, or simply the octonion KdV equation, is given by

$$u_t + u_{xxx} + \frac{1}{2}(u^2)_x = 0, \tag{98}$$

when $\vec{B} = \vec{0}$ it reduces to the scalar KdV equation. In terms of b and $\vec{B}$ the equation can be reexpressed as

$$b_t + b_{xxx} + bb_x - \sum_{i=1}^{7} B_i B_{ix} = 0, \tag{99}$$

$$(B_i)_t + (B_i)_{xxx} + (bB_i)_x = 0. \tag{100}$$

Eq. (98) is invariant under the Galileo transformation given by

$$\begin{aligned} \tilde{x} &= x + ct, \\ \tilde{t} &= t, \\ \tilde{u} &= u + c \end{aligned} \tag{101}$$

where c is a real constant.

Additionally, Eq. (98) is invariant under the automorphisms of the octonions, that is, under the group G_2. If under an automorphism

$$u \to \phi(u) \tag{102}$$

then

$$u_1 u_2 \to \phi(u_1 u_2) = \phi(u_1)\phi(u_2) \tag{103}$$

http://dx.doi.org/10.5772/66146

and consequently

$$[\phi(u)]_t + [\phi(u)]_{xxx} + \frac{1}{2}\left([\phi(u)]^2\right)_x = 0. \tag{104}$$

12. The Gardner formulation for the octonion valued algebra KdV equation

Associated with the real KdV equation, there is a Gardner ε-transformation and a Gardner equation which allows to obtain in a direct way the corresponding infinite sequence of conserved quantities. There exists a generalization of this approach for the KdV valued on the octonion algebra. The generalized Gardner transformation, expressed in terms of a new field $r(x, t)$ valued on the octonion is given by

$$u = r + \varepsilon r_x - \frac{1}{6}\varepsilon^2 r^2. \tag{105}$$

The generalized Gardner equation is then

$$r_t + r_{xxx} + \frac{1}{2}(r r_x + r_x r) - \frac{1}{12}\left((r^2) r_x + r_x (r^2)\right)\varepsilon^2 = 0 \tag{106}$$

where ε is a real parameter.

If $r(x, t)$ is a solution of the generalized Gardner equation (106), then $u(x, t)$ is a solution of the octonion algebra valued KdV equation (98). It has been shown in Ref. [23] that $\int_{-\infty}^{+\infty} \mathbb{R}e[r(x, t)]dx$ is a conserved quantity of Eq. (106). We can then invert Eq. (105), assuming a formal ε-expansion of the solution $r(x, t)$, to obtain an infinite sequence of conserved quantities for the KdV equation valued on the octonion algebra.

13. The master Lagrangian for the KdV equation valued on the octonion algebra

We may now use the Helmholtz procedure to obtain a Lagrangian density for the generalized Gardner equation. The master Lagrangian formulated in terms of the Casimir potential $s(x, t)$,

$$r(x, t) = s_x(x, t), \tag{107}$$

is

$$L_\varepsilon(s) = \int_{t_i}^{t_f} dt \int_{-\infty}^{+\infty} \mathcal{L}_\varepsilon(s) dx \tag{108}$$

where the Lagrangian density is given by

$$\mathcal{L}_\varepsilon(s) = \mathbb{R}e\left[-\frac{1}{2}s_x s_t - \frac{1}{6}(s_x)^3 + \frac{1}{2}(s_{xx})^2 + \frac{1}{72}\varepsilon^2(s_x)^4\right]. \tag{109}$$

The Lagrangian density $\mathcal{L}_\varepsilon(s)$ is invariant under the action of the exceptional Lie group G_2.

Independent variations with respect to s yields

$$\begin{aligned}\delta\mathcal{L}_\varepsilon(s) &= \mathbb{R}e\left[-\frac{1}{2}(\delta s)_x s_t - \frac{1}{2}s_x(\delta s)_t - \frac{1}{6}\left((\delta s)_x(s_x)^2 + s_x(\delta s)_x s_x + (s_x)^2(\delta s)_x\right)\right] \\ &+ \mathbb{R}e\left[\frac{1}{2}\left((\delta s)_{xx}s_{xx} + (\delta s)_{xx}s_{xx}\right) + \frac{1}{72}\varepsilon^2\left((\delta s)_x(s_x)^3 + s_x(\delta s)_x(s_x)^2 + (s_x)^2(\delta s)_x s_x + (s_x)^3(\delta s)_x\right)\right].\end{aligned} \tag{110}$$

Using properties of the octonion algebra we obtain from the stationary requirement $\delta L_\varepsilon(s) = 0$ the generalized Gardner equation (106).

In the calculation the property to be a division algebra of the octonions is explicitly used.

If we take the limit $\epsilon \to 0$, we obtain a first Lagrangian for the KdV equation valued on the octonion algebra,

$$L(w) = \int_{t_i}^{t_f} dt \int_{-\infty}^{+\infty} dx \mathbb{R}e\left[-\frac{1}{2}w_x w_t - \frac{1}{6}(w_x)^3 + \frac{1}{2}(w_{xx})^2\right]. \tag{111}$$

Independent variations with respect to w yields, using $u = w_x$, the octonionic KdV equation (98). If we consider the following redefinition

$$\begin{aligned} s \to \hat{s} &= \varepsilon s \\ \mathcal{L}_\varepsilon(s) &\to \varepsilon^2\mathcal{L}_\varepsilon(\hat{s}) \end{aligned} \tag{112}$$

and take the limit $\epsilon \to \infty$ we obtain

$$\lim_{\epsilon\to\infty} \epsilon^2\mathcal{L}_\epsilon(\hat{s}) = \mathcal{L}^M(\hat{s}), \tag{113}$$

where

$$\mathcal{L}^M(\hat{s}) = \mathbb{R}e\left[-\frac{1}{2}\hat{s}_x\hat{s}_t + \frac{1}{2}(\hat{s}_{xx})^2 + \frac{1}{72}(\hat{s}_x)^4\right]. \tag{114}$$

We get in this limit the generalized Miura Lagrangian

$$L^M(\hat{s}) = \int_{t_i}^{t_f} dt \int_{-\infty}^{+\infty} dx \mathcal{L}^M(\hat{s}). \tag{115}$$

The Miura equation is then obtained by taking variations with respect to $\hat{s}$, we get

$$\hat{r}_t + \hat{r}_{xxx} - \frac{1}{18}(\hat{r})_x^3 = 0, \quad \hat{r} \equiv \hat{s}_x, \tag{116}$$

http://dx.doi.org/10.5772/66146

while the Miura transformation arises after the redefinition process, it is $u = \hat{r}_x - \frac{1}{6}\hat{r}^2$.

Any solution of the Miura equation, through the Miura transformation, yields a solution of the KdV equation valued on the octonion algebra. Since $L_\varepsilon(s)$ is invariant under G_2, the same occurs for $L(w)$ and $L^M(\hat{s})$, and consequently for the equations arising from variations of them.

The Lagrangian formulation of the octonionic KdV equation may be used as the starting step to obtain the Hamiltonian structure of the octonion algebra valued KdV equation.

14. Conclusions

We analyzed the relevance of the Dirac approach for constraint systems applied to singular Lagrangians. Several interesting theories are described by singular Lagrangians, notoriously the gauge theories describing the known fundamental forces in nature. In this chapter, we emphasized its relevance in the formulation of completely integrable field theories. We discussed extensions of the Korteweg-de Vries equation in different contexts. All these extensions, together with the KdV equation, allow a construction of a Lagrangian and a Hamiltonian structure arising from the application of the Helmholtz procedure. That is, starting with a time evolution partial differential system we construct, following the Helmholtz procedure, a Lagrangian associated with it. We present the construction of several Lagrangians and their corresponding Hamiltonian structures associated with the coupled KdV systems. All of them are characterized by second class constraints. The physical phase space is obtained by the determination of the complete set of constraints and the corresponding Dirac brackets. We established the relation between the several constructions by obtaining a pencil of Poisson structures. The application includes systems with an infinite sequence of conserved quantities together with a system with finite number of conserved quantities but presenting soliton solutions with nice stability properties. The final application is an extension of the KdV equation to the case in which the fields are valued on the octonion algebra. We constructed a master formulation from which two dual Lagrangian formulations are obtained , one corresponding to the KdV valued on the octonions and the other one corresponding to the extension of the modified KdV equation to fields valued on the octonions.

One important extrapolation of the analysis we have presented is the construction of gauge theories describing completely integrable systems. In fact, it is natural to extend the analysis by constructing a gauge theory which under a gauge fixing procedure reduces to the completely integrable systems of the KdV type we have discussed.

Acknowledgements

A. R. and A. S. are partially supported by Project Fondecyt 1161192, Chile.

Pacs: 02.30.lk, 11.30.-j, 02.30. Jr, 02.10.Hh

Author details

Alvaro Restuccia[1,2]* and Adrián Sotomayor[3]*

*Address all correspondence to: arestu@usb.ve and adrian.sotomayor@uantof.cl

1 Departamento de Física, Universidad de Antofagasta, Antofagasta, Chile

2 Departamento de Física, Universidad Simón Bolívar, Caracas, Venezuela

3 Departamento de Matemáticas, Universidad de Antofagasta, Antofagasta, Chile

References

[1] C. S. Gardner, J. M. Greene, M. D. Kruskal and R. M. Miura, Phys. Rev. Lett. 19, 1096 (1967).

[2] R. M. Miura, J. Math. Phys. 9, 1202 (1968).

[3] R. M. Miura, C. S. Gardner and M. D. Kruskal, J. Math. Phys. 9, 1204 (1968).

[4] C. Gardner, J. Math. Phys. 12, 1548 (1971).

[5] R. M. Miura, Siam Rev. 18, 412 (1976).

[6] R. Hirota and J. Satsuma, Phys. Lett. 85A, 407 (1981).

[7] P. J. Olver and V. V. Sokolov, Commun. Math. Phys. 193, 245 (1998).

[8] B. A. Kupershmidt, Phys. Lett. A 102, 449 (1984).

[9] P. Mathieu, J. Math. Phys. 29, 2499 (1988).

[10] P. Labelle and P. Mathieu, J. Math. Phys. 32, 923 (1991).

[11] S. Belucci, E. Ivanov and S. Krivonos, J. Math. Phys. 34, 3087 (1993).

[12] F. Delduc and E. Ivanov, Phys. Lett. B 309, 312–319 (1993).

[13] F. Delduc, E. Ivanov and S. Krivonos, J. Math. Phys. 37, 1356 (1996).

[14] Z. Popowicz, Phys. Lett. B 459, 150–158 (1999).

[15] A. Restuccia and A. Sotomayor, J. Math. Phys. 54, 113510 (2013).

[16] S. Andrea, A. Restuccia and A. Sotomayor, J. Math. Phys. 42, 2625 (2001).

[17] J. A. Gear and R. Grimshaw, Stud. Appl. Math. 70, 235 (1984); J. A. Gear, Stud. Appl. Math. 72, 95 (1985).

[18] S. Y. Lou, B. Tong, H. C. Hu and X. Y. Tang, J. Phys. A: Math. Gen. 39, 513–527 (2006).

[19] V. A. Brazhnyi and V. V. Konotop, Phys. Rev. E 72, 026616 (2005).

[20] P. A. M. Dirac, "Lectures on Quantum Mechanics," Belfer Graduate School Monograph Series No.2, Yeshiva University, New York (1964).

[21] L. C. Vega, A. Restuccia and A. Sotomayor, Contribution to the Proceedings of the 2nd International Conference on Mathematical Modeling in Physical Sciences 2013, J. Phys. Conf. Series 490, 012024 (2014).

[22] A. Restuccia and A. Sotomayor, Open Phys. 14, 95–105 (2016).

[23] A. Restuccia, A. Sotomayor and J. P. Veiro, arXiv: 1609.05410 [math-ph] (2016).

[24] J. L. Bona, S. Vento and F. B. Weissler, Discrete Contin. Dyn. Sys. 33, 11–12, 4811 (2013).

[25] J. R. Yang and J. J. Mao, Commun. Theor. Phys. 49, 22–26 (2008).

[26] R. Hirota, Phys. Rev. Lett. 27(18), 1192–1194 (1971).

[27] L. C. Vega, A. Restuccia and A. Sotomayor, arXiv:1407.7743 v3 [math-ph].

[28] H. D. Wahlquist and F. B. Estabrook, Phys. Rev. Lett. 31, 1387 (1973).

[29] S. Yu. Sakovich, J. Nonlin. Math. Phys. 6(3), 255–262 (1999).

[30] A. K. Karasu, J. Math. Phys. 38(7), 3616–3622 (1997).

[31] P. Casati and G. Ortenzi, J. Geom. and Phys. 56, 418–449 (2006).

[32] S. Andrea, A. Restuccia and A. Sotomayor, Phys. Lett. A 376, 245–251 (2012).

[33] P. Dargis and P. Mathieu, Phys. Lett. A 176, 67–74 (1993).

[34] A. Restuccia and A. Sotomayor, Boundary Value Problems 2013, 224 (2013).

[35] T. B. Benjamin, Proc. R. Soc. Lond. A 328, 153–183 (1972).

[36] J. M. Evans, Nucl. Phys. B 298, 92–108 (1988).

Chapter 2

Lagrangian Subspaces of Manifolds

Yang Liu

Additional information is available at the end of the chapter

http://dx.doi.org/ 10.5772/67290

Abstract

In this chapter, we provide an overview on the Lagrangian subspaces of manifolds, including but not limited to, linear vector spaces, Riemannian manifolds, Finsler manifolds, and so on. There are also some new results developed in this chapter, such as finding the Lagrangians of complex spaces and providing new insights on the formula for measuring length, area, and volume in integral geometry. As an application, the symplectic structure determined by the Kähler form can be used to determine the symplectic form of the complex Holmes-Thompson volumes restricted on complex lines in integral geometry of complex Finsler space. Moreover, we show that the space of oriented lines and the tangent bundle of unit sphere in Minkowski space are symplectomorphic.

Keywords: Lagrangian subspace, differential geometry

1. Introduction

In differential geometry and differential topology, manifolds are the main objects being studied, and Lagrangian submanifolds are submanifolds that carry differential forms with special property, which are usually called symplectic form in real manifolds and Kahler form in complex manifolds.

This book chapter is concerned with explicit canonical symplectic form for real and complex spaces and answer to the questions on the existence of Lagrangian subspace. One can find and explicitly describe the set of Lagrangian subspaces of $\mathbb{R}^2$ with L^p norm, $1 \leq p < \infty$, as a an example of Finsler spaces. Since Holmes-Thompson volumes, as measures, depend on the differential structures of the spaces, the symplectic structure determined by the symplectic form can be used to determine the symplectic form of Holmes-Thompson volumes restricted on lines in integral geometry of L^p spaces, as an application to integral geometry.

Some ingenuous ideas in physics and engineering actually originated from mathematics. For example, the relativity theory in physics, to some sense, originated from Riemmanian geometry.

The real Finsler spaces, as generalizations of real Riemannian manifolds, were introduced in Ref. [1] about a century ago and have been studied by many researchers (see, for instance, Refs. [2–4]), and Finsler spaces (see, for instance, Refs. [5, 6]) have become an interest of research for the studies of geometry, including differential geometry and integral geometry, in recent decades. By the way, there are applications of Finsler geometry in physics and engineering, and in particular, Finsler geometry can be applied to engineering dynamical systems, on which one can see Ref. [7]. As a typical Finsler space, L^p space, $1 < p < \infty$, has the main features of a Finsler space. As such, we focus on L^p space, $1 < p < \infty$, in this chapter, but some results can be generalized to general Finsler spaces, on which one can refer to Ref. [8]. The L^p space, $1 < p < \infty$, as a generalization of Euclidean space, has a rich structure in functional analysis (see, for instance, Refs. [9, 10]), and particularly in Banach space. Furthermore, it has broad applications in statistics (see, for instance, Refs. [11, 12]), engineering (see, for instance, Ref. [13, 27]), mechanics (see, for instance, Ref. [14]), computational science (see, for instance, Ref. [15]), biology (see, for instance, Ref. [16]), and other areas. Along this direction, L^p, $0 < p \leq 1$, in the sense of conjugacy to the scenario of L^p, $1 < p < \infty$, also has broad applications, in particular, signal processing in engineering, on which one can see Refs. [17–19].

This chapter is structured as follows: In Section 2, we provide a description on Gelfand transform, which is one of the most fundamental transforms in integral geometry; in Section 3, we introduce density needed for the measure of length of curves; in Section 4, we further study the Lagrangian subspaces of complex L^p spaces; in Section 5, we work on tangent bundle of unit sphere in Minkowski space and its symplectic or Lagrangian structure; in Section 6, we apply the Lagrangian structure to establish the length formula in integral geometry; and in Section 7, we further apply the Lagrangian structure of a Minkowski space to establish the formula for the Holmes-Thompson area in integral geometry.

2. Gelfand transform

Given a double fibration:

$$\mathbb{R}^2 \overset{\pi_1}{\leftarrow} \mathrm{F} \overset{\pi_2}{\rightarrow} \overline{Gr_1(\mathbb{R}^2)} \tag{1}$$

where

$$\begin{aligned} \mathrm{F} = &\left\{((x,y), l(r,\theta)) : (x,y) \in \mathbf{R}^2, l(r,\theta) \in \overline{Gr_1(\mathbf{R}^2)}, (x,y) \in l(r,\theta)\right\} \\ &\simeq \left\{(x,y,r,\theta) : x\cos(\theta) + y\sin(\theta) = r\right\}, \end{aligned}$$

π_1 and π_2 are the natural projections of fibers. The Gelfand transform of a 2-density $\varphi = |dr \wedge d\theta|$ is defined as

$$\mathrm{GT}(\varphi) = \pi_{1_*}\pi_2^*\varphi, \tag{2}$$

which is a 1-density $\mathbb{R}^2$.

http://dx.doi.org/ 10.5772/67290

3. 1-Density

Lemma 3.1. For any $v = (\alpha,\beta) \in T_{(x,y)}\mathbb{R}^2$,

$$\mathrm{GT}(\varphi)((x,y),\ v) = 4|v|. \tag{3}$$

Proof. For $v = (\alpha,\beta) \in T_{(x,y)}\mathbf{R}^2$, there exists

$$\tilde{v} = (\alpha,\beta,\alpha\cos(\theta) + \beta\sin(\theta), \theta) \in T_{((x,y),\ l(r,\theta))}\mathrm{F}, \tag{4}$$

such that $\mathrm{d}\pi_1(\tilde{v}) = v$. Therefore, we have

$$\begin{aligned}
\mathrm{GT}(\varphi)(v) &= \int\limits_{\pi_1^{-1}((x,y))} \pi_2^*\varphi(\tilde{v},\bullet) \\
&= \int\limits_{\{(x,y),l(p,\theta):\ x\cos(\theta)+y\sin(\theta)=r\}} |\mathrm{d}r\wedge\mathrm{d}\theta|(\tilde{v},\bullet) \\
&= \int_0^{2\pi} |\alpha\cos(\theta) + \beta\sin(\theta)|d\theta \\
&= \int_0^{2\pi} |v\cdot(\cos(\theta),\sin(\theta))|d\theta \\
&= |v|\int_0^{2\pi} \cos(\theta_0+\theta)\mathrm{d}\theta \text{ where } \alpha = |v|\cos(\theta_0), \beta = |v|\sin(\theta_0) \\
&= 4|v|.
\end{aligned} \tag{5}$$

Remark 3.2. By Alvarez's Gelfand transform for Crofton type formulas, we know that

$$\int_{l\in\mathbf{R}^2} \#(\gamma\cap l(r,\theta))\mathrm{d}r\mathrm{d}\theta = \int_\gamma \mathrm{GT}(\varphi). \tag{6}$$

Thus, we have now proved the Crofton formula: Given a differentiable curve γ in $\mathbb{R}^2$, the length of γ can be computed in the following formula:

$$\mathrm{Length}(\gamma) = \frac{1}{4}\int_{l\in\mathbf{R}^2} \#(\gamma\cap l(r,\theta))\mathrm{d}r\mathrm{d}\theta. \tag{7}$$

4. Lagrangian subspaces of complex spaces

Some of the results have obtained in Ref. [8], but because the Lagrangian subspaces of complex spaces are essential to establish the generalized volume formula in complex integral geometry, let us give an expository on the Kahler strut rue of generalized complex spaces.

Theorem 4.1. *The set of Lagrangian subspaces of* $\mathbb{C}^2$ *with* L^1 *norm is* $\mathbb{T}^2 \cup \mathbb{T}^1$, *where*

$$\mathbb{T}^2 := \{\text{span}((z,0),(0,w)) : z,w \in U(1)\} \cong U(1) \times U(1) \tag{8}$$

and

$$\mathbb{T}^1 := \{P : P = \{\lambda(z,w) : \lambda \in \mathbb{R}, z,w \in U(1),\ zw \text{ is a constant in } U(1)\}\} \cong U(1). \tag{9}$$

Proof. First, we can show that

$$P = \{\lambda(z,w) : \lambda \in \mathbb{R}, z,w \in U(1),\ zw \text{ is a constant in } U(1)\} \tag{10}$$

is identical to some

$$P' := \text{span}((z_1, z_1 e^{i\theta}), (z_2, \frac{z_1^2 \overline{z_2}}{|z_1|^2} e^{i\theta})) \tag{11}$$

where $z_1, z_2 \in \mathbb{C} \backslash \{0\}$. For any $\lambda(e^{i\varphi}, e^{i\psi}) \in P$, let $z_1 = \lambda e^{i\varphi}$, $\theta = \psi - \varphi$, we have $P = \text{span}((z_1, z_1 e^{i\theta}),$ $(z_2, \frac{z_1^2\overline{z_2}}{|z_1|^2} e^{i\theta})) = P'$ where $z_2 \in \mathbb{C}\backslash\{0\}$.

We can get $\kappa_1(z_1,0),(0,z_2)) = 0$. On the other hand, for any

$$(z,w) = \lambda_1(z_1,z_1) + \lambda_2(z_2, \frac{z_1^2\overline{z_2}}{|z_1|^2}) \in \text{span}((z_1,z_1),(z_2,\frac{z_1^2\overline{z_2}}{|z_1|^2})), \tag{12}$$

where $\lambda_1, \lambda_2 \in \mathbb{R}$,

$$\begin{aligned} |w|^2 &= (\lambda_1 z_1 + \lambda_2 \frac{z_1^2\overline{z_2}}{|z_1|^2})(\lambda_1 \overline{z_1} + \lambda_2 \frac{\overline{z_1}^2 z_2}{|z_1|^2}) \\ &= \lambda_1^2 z_1 \overline{z_1} + \lambda_1\lambda_2 \overline{z_1} z_2 + \lambda_2 \lambda_1 z_1 \overline{z_2} + \lambda_2^2 \overline{z_2} z_2 \\ &= (\lambda_1 z_1 + \lambda_2 z_2)(\lambda_1 \overline{z_1} + \lambda_2 \overline{z_2}) \\ &= |z|^2, \end{aligned} \tag{13}$$

that implies $|\frac{w}{z}| = 1$. Therefore, we have

$$\begin{aligned} \kappa_{(z,w)}((z_1,z_1),(z_2,\tfrac{z_1^2\overline{z_2}}{|z_1|^2})) &= \tfrac{3}{2}(Im(z_2\overline{z_1}) + \tfrac{3}{2} Im(\tfrac{z_1^2\overline{z_2}}{|z_1|^2}\overline{z_1})) \\ &\quad -\tfrac{1}{2} Im(\tfrac{z}{w}|\tfrac{w}{z}|(\tfrac{z_1^2\overline{z_2}}{|z_1|^2}\overline{z_1} - z_1\overline{z_2})) \\ &= \tfrac{3}{2}(Im(z_2\overline{z_1}) + Im(z_1\overline{z_2})) \\ &= 0. \end{aligned} \tag{14}$$

So κ vanishes on $\text{span}((z_1,z_1),(z_2,\frac{z_1^2\overline{z_2}}{|z_1|^2}))$ for any $z_1,z_2 \in \mathbb{C}\backslash\{0\}$, $Im(\frac{z_1}{\overline{z_1}}) \neq 0$.

Conversely, suppose that κ vanishes on a plane P spanned by (z_1,w_1) and (z_2,w_2). We know that

$$(1+\frac{1}{2}|\frac{w}{z}|)Im(z_2\overline{z_1})+(1+\frac{1}{2}|\frac{z}{w}|)Im(w_2\overline{w_1})+\frac{1}{2}Im(\frac{z}{w}|\frac{w}{z}|(w_2\overline{z_1}-w_1\overline{z_2}))=0 \tag{15}$$

holds for any $(z,w)\in\text{span}((z_1,w_1),(z_2,w_2))$. In the following argument, we divide it into three cases to discuss in terms of $|\frac{w}{z}|$ and $\frac{w}{z}|\frac{w}{z}|$.

The first case is that $|\frac{w}{z}|=\lambda$ for some fixed $\lambda>0$. Let $(z,w)=\lambda_1(z_1,w_1)+\lambda_2(z_2,w_2)$ for any $\lambda_1,\lambda_2\in\mathbb{R}$, then $|\lambda_1 w_1+\lambda_2 w_2|=\lambda|\lambda_1 z_1+\lambda_2 z_2|$, that implies $|w_1|=\lambda|z_1|$, $|w_2|=\lambda|z_2|$ and $Re(w_1\overline{w_2})=\lambda^2 Re(z_1\overline{z_2})$. It follows that $w_1=\lambda e^{i\theta}z_1$, $w_2=\lambda e^{i\theta}z_2$, or $w_1=\lambda e^{i\theta}z_1$, $w_2=\lambda e^{i\theta}\frac{z_1^2\overline{z_2}}{|z_1|^2}$ for some $\theta\in[0,2\pi)$.

In the sub-case of $w_1=\lambda e^{i\theta}z_1$, $w_2=\lambda e^{i\theta}z_2$ for some $\theta\in[0,2\pi)$, by Eq. (**15**) we have

$$(1+\frac{\lambda}{2})Im(z_2\overline{z_1})+(1+\frac{1}{2\lambda})\lambda^2 Im(z_2\overline{z_1})+\lambda Im(z_2\overline{z_1})=(1+\lambda)^2 Im(z_2\overline{z_1})=0, \tag{16}$$

which implies $Im(z_2\overline{z_1})=0$ and furthermore $Im(w_2\overline{w_1})=0$. That means (z_1,w_1) and (z_2,w_2) are colinear. So this case cannot occur.

However, for the other sub-case of $w_1=\lambda e^{i\theta}z_1$, $w_2=\lambda e^{i\theta}\frac{z_1^2\overline{z_2}}{|z_1|^2}$ for some $\theta\in[0,2\pi)$, by Eq. (**15**) we have

$$(1+\frac{\lambda}{2})Im(z_2\overline{z_1})+(1+\frac{1}{2\lambda})\lambda^2 Im(z_1\overline{z_2})=(1-\lambda^2)Im(z_2\overline{z_1})=0. \tag{17}$$

Then $\lambda=1$ or $Im(z_2\overline{z_1})=0$, but (z_1,w_1) and (z_2,w_2) cannot be colinear. So, we have $\lambda=1$ which gives

$$P=\text{span}((z_1,z_1e^{i\theta}),(z_2,\frac{z_1^2\overline{z_2}}{|z_1|^2}e^{i\theta})), \tag{18}$$

where $z_1,z_2\in\mathbb{C}\backslash\{0\}$ and $Im(z_1\overline{z_2})\neq0$ for some $\theta\in[0,2\pi)$. This finishes the first case.

The second case is $\frac{w}{z}|\frac{w}{z}|=e^{i\theta}$ for some fixed $\theta\in[0,2\pi)$. Let $w_1=\lambda_1e^{i\theta}z_1, w_2=\lambda_2e^{i\theta}z_2$ for some $\lambda_1,\lambda_2>0$. Then it follows from (15) that

$$\begin{array}{rl} & (1+\frac{\lambda_1}{2})Im(z_2\overline{z_1})+(1+\frac{1}{2\lambda_1})\lambda_1\lambda_2 Im(z_2\overline{z_1})+\frac{1}{2}(\lambda_1+\lambda_2)Im(z_2\overline{z_1})\\ = & (1+\frac{\lambda_2}{2})Im(z_2\overline{z_1})+(1+\frac{1}{2\lambda_2})\lambda_1\lambda_2 Im(z_2\overline{z_1})+\frac{1}{2}(\lambda_1+\lambda_2)Im(z_2\overline{z_1})\\ = & (1+\lambda_1)(1+\lambda_2)Im(z_2\overline{z_1})\\ = & 0\end{array} \tag{19}$$

at the points (z_1,w_1) and (z_2,w_2), which implies $Im(z_2\overline{z_1})=0$ and furthermore $Im(w_2\overline{w_1})=0$. Thus, z_1 and z_2, w_1, and w_2 are colinear, which implies that P equals a plane spanned by one vector from $\{(z_1,0),(z_2,0)\}$ and the other from $\{(0,w_1),(0,w_2)\}$. Thus $P\in\mathbb{T}^2$.

The last case is the negative to the first one and the second one. It gives $Im(z_2\overline{z_1})=Im(w_2\overline{w_1})$ $=0$ and $w_2\overline{z_1}-w_1\overline{z_2}=0$ because of the linear independence, but the former implies the latter by linear transformation, so it is brought down to $Im(z_2\overline{z_1}) = Im(w_2\overline{w_1}) = 0$. Thus, we have $P \in \mathbb{T}^2$ by the second case, and that concludes the proof.

5. Tangent bundle of uni-sphere in Minkowski space and symplectic or Lagrangian structure

In this section, we show that the space of oriented lines and the tangent bundle of unit sphere in Minkowski space are symplectomorphic.

Let us consider a Minkowski plane $(\mathbb{R}^2, F)$ first, where F is a Finsler metric. The natural symplectic form on $T^*\mathbb{R}^2$ is $dx \wedge d\overline{\xi} + dy \wedge d\overline{\eta}$, and then the natural symplectic form on $T\mathbb{R}^2$ induce by the Finsler metric F is

$$\begin{aligned} \omega &:= dx \wedge d\frac{\partial F}{\partial \xi} + dy \wedge d\frac{\partial F}{\partial \eta} \\ &= \frac{\partial^2 F}{\partial \xi^2} dx \wedge d\xi + \frac{\partial^2 F}{\partial \xi \partial \eta}(dx \wedge d\eta + dy \wedge d\xi) + \frac{\partial^2 F}{\partial \eta^2} dy \wedge d\eta. \end{aligned} \tag{20}$$

Define a projection $\pi : T\mathbb{R}^2 \to \overline{Gr_1(\mathbb{R}^2)}$ by

$$\pi((x,y);(\xi,\eta)) = ((x,y) - dF(\xi,\eta)((x,y))(\xi,\eta);(\xi,\eta)). \tag{21}$$

Let S_F be the unit circle in the Minkowski plane and TS_F be its tangent bundle. It is a fact that $TS_F \cong \overline{Gr_1(\mathbb{R}^2)}$. On the other hand, since TS_F is embedded in $T\mathbb{R}^2$, it inherits a natural symplectic form $\omega_0 := \omega|_{TS_F}$ from $T\mathbb{R}^2$.

Theorem 5.1. $\pi^*\omega_0 = \omega|_{S^*\mathbb{R}^2}$.

Proof. Applying the equality

$$\frac{\partial F}{\partial \xi} d\xi + \frac{\partial F}{\partial \eta} d\eta = 0, \tag{22}$$

we obtain

$$\begin{aligned} \pi^*\omega_0 = &\frac{\partial^2 F}{\partial \xi^2} d(x - dF(\xi,\eta)((x,y))\xi) \wedge d\xi + \frac{\partial^2 F}{\partial \xi \partial \eta}(d(x - dF(\xi,\eta)((x,y))\xi) \wedge d\eta \\ &+ d(y - dF(\xi,\eta)((x,y))\eta) \wedge d\xi) + \frac{\partial^2 F}{\partial \eta^2} d(y - dF(\xi,\eta)((x,y))\eta) \wedge d\eta \\ = &\frac{\partial^2 F}{\partial \xi^2} dx \wedge d\xi + \frac{\partial^2 F}{\partial \xi \partial \eta}(dx \wedge d\eta + dy \wedge d\xi) + \frac{\partial^2 F}{\partial \eta^2} dy \wedge d\eta \\ &- d(dF(\xi,\eta)((x,y))) \wedge (\frac{\partial^2 F}{\partial \xi^2} \xi d\xi + \frac{\partial^2 F}{\partial \eta^2} \eta d\eta + \frac{\partial^2 F}{\partial \xi \partial \eta}(\xi d\eta + \eta d\xi)). \end{aligned} \tag{23}$$

By the positive homogeneity of F, one can get the useful fact that $F(\xi,\eta) = \xi\frac{\partial F}{\partial \xi} + \eta\frac{\partial F}{\partial \eta}$. Therefore,

$$\xi\frac{\partial F}{\partial \xi} + \eta\frac{\partial F}{\partial \eta} = 1. \tag{24}$$

By differentiating (24), we get

$$\frac{\partial^2 F}{\partial \xi^2}\xi d\xi + \frac{\partial^2 F}{\partial \eta^2}\eta d\eta + \frac{\partial^2 F}{\partial \xi \partial \eta}(\xi d\eta + \eta d\xi) + \frac{\partial F}{\partial \xi}d\xi + \frac{\partial F}{\partial \eta}d\eta = 0. \tag{25}$$

Applying (22) again, we have

$$\frac{\partial^2 F}{\partial \xi^2}\xi d\xi + \frac{\partial^2 F}{\partial \eta^2}\eta d\eta + \frac{\partial^2 F}{\partial \xi \partial \eta}(\xi d\eta + \eta d\xi) = 0. \tag{26}$$

Thus, the claim follows.

Remark 5.2. For a n-dimensional Minkowski space $(\mathbb{R}^n,F)$, we just need to add more indices, then the theorem above is also true for $(\mathbb{R}^n,F)$.

Therefore, letting F be a Finsler metric on $\mathbb{R}^n$ and S_F be the unit sphere in the Minkowski space $(\mathbb{R}^n,F)$, we obtain the following general theorem:

Theorem 5.3. *The symplectic form on the space of lines in a Minkowski space $(\mathbb{R}^n,F)$ is the canonical symplectic form on the tangent bundle TS_F as imbedded in $T\mathbb{R}^n$.*

We have the following remarks:

Remark 5.4. Theorem 5.3 provides a perspective that we can transform calculus on $\overline{Gr_1(\mathbb{R}^2)}$ to ones on TS_F.

and

Remark 5.5. We can analyze the differential structure of the Minkowski space by considering its symplectic form or Lagrangian structure. The Lagrangian structure of tangent spaces of Minkowski space gives the symplectic structure of the space of geodesics in the Minkowski space, and in general, the measures on a space or manifold in integral geometry depend on the differential structures of the space or manifold. Holmes-Thompson volumes are defined based on Lagrangian structure (see, for instance, Refs. [12, 20]), so, as an application, the symplectic structure determined by the symplectic form can be used to determine the symplectic form of the Holmes-Thompson volumes restricted on lines in integral geometry of Minkowski space, about which one can see Refs. [21–23].

Another remark from the proof of Theorem 5.1 is that

Remark 5.6. A combination of (26) and Gelfand transform (see Ref. [6]) may be used to provide a short proof of the general Crofton formula for Minkowski space.

6. Application to generalized length and related

For any rectifiable curve γ in the Euclidean plane, the classic Crofton formula is

$$\text{Length}(\gamma) = \frac{1}{4}\int_0^{\infty}\int_0^{2\pi} \#(\gamma \cap l(r,\theta))\mathrm{d}\theta \mathrm{d}r, \tag{27}$$

where θ is the angle from the x-axis to the normal of the oriented line l and r is the distance form the origin to l. Let us denote the affine 1-Grassmannians consisting of lines in $\mathbb{R}^2$ by $\overline{Gr_1(\mathbb{R}^2)}$.

As for Minkowski plane, it is a normed two dimensional space with a norm $F(\cdot) = ||\cdot||$, in which the unit disk is convex and F has some smoothness.

Two significant and useful tools that are used to obtain the Crofton formula for Minkowski plane are the cosine transform and Gelfand transform. Let us explain them one by one first and see the connections between them later. A important fact or result from spherical harmonics about cosine transform is that there is some even function on S^1 such that

$$F(\cdot) = \int_{S^1} |\langle \xi, \cdot \rangle| g(\theta)\mathrm{d}\theta, \tag{28}$$

if F is an even C^4 function on S^1. A great reference for this would be [24] by Groemer. As for Gelfand transform, it is the transform of differential forms and densities on double fibrations, for instance, $\mathbb{R}^2 \overset{\pi_1}{\leftarrow} \mathcal{I} \overset{\pi_2}{\rightarrow} \overline{Gr_1(\mathbb{R}^2)}$, where $\mathcal{I} := \left\{(x,l) \in \mathbb{R}^2 \times \overline{Gr_1(\mathbb{R}^2)} : x \in l\right\}$ is the incidence relations and π_1 and π_2 are projections. A formula one can take as an example of the fundamental theorem of Gelfand transform is the following:

$$\int_{\gamma} \pi_{1*}\pi_2^*|\Omega| = \int_{l \in \overline{Gr_1(\mathbb{R}^2)}} \#(\gamma \cap l)|\Omega|, \tag{29}$$

where $\Omega := g(\theta)\mathrm{d}\theta \wedge \mathrm{d}r$. However, here we provide a direct proof for this fundamental theorem of Gelfand transform.

Proof. First, consider the case of $\Omega = \mathrm{d}\theta \wedge \mathrm{d}r$. For any $v \in T_x\gamma$, since there is some $v' \in T_{x'}\mathcal{I}$, such that $\pi_{1*}(v') = v$, then

$$\begin{aligned} (\pi_{1*}\pi_2^*|\Omega|)_x(v) &= (\int_{\pi_1^{-1}(x)} \pi_2^*|\Omega|)_x(v) \\ &= \int_{x' \in \pi_1^{-1}(x)} (\pi_2^*|\Omega|)_{x'}(v') \\ &= \int_{S^1} (\pi_2^*|\mathrm{d}\theta \wedge \mathrm{d}r|)(v') \\ &= \int_{S^1} |\mathrm{d}r(\pi_{2*}(v'))|\mathrm{d}\theta \\ &= \int_{S^1} |\langle v,\theta \rangle|\mathrm{d}\theta \\ &= 4|v|. \end{aligned} \tag{30}$$

Thus, we have

$$\int_{\gamma} \pi_{1*}\pi_2^*|\Omega| = 4\text{Length}(\gamma) = \int_{l\in\overline{Gr_1(\mathbb{R}^2)}} \#(\gamma\cap l)|\Omega| \tag{31}$$

by using the classic Crofton formula.

For the general case of $\Omega = f(\theta)\mathrm{d}\theta\wedge\mathrm{d}r$, we just need to substitute $\mathrm{d}\theta$ by $g(\theta)\mathrm{d}\theta$ in the equalities in the first case.

Furthermore, we can also see, from the above proof and eq:exist, that

$$\int_{\gamma} \pi_{1*}\pi_2^*|\Omega| = \int_a^b (\pi_{1*}\pi_2^*|\Omega|)(\gamma'(t))\mathrm{d}t = \int_a^b 4F(\gamma'(t))\mathrm{d}t = 4\text{Length}(\gamma), \tag{32}$$

for any curve $\gamma(t) : [a,b] \to \mathbb{R}^2$ differentiable almost everywhere in the Minkowski space. Therefore, by using (29), we obtain that

$$\text{Length}(\gamma) = \frac{1}{4}\int_{l\in\overline{Gr_1(\mathbb{R}^2)}} \#(\gamma\cap l)|g(\theta)\mathrm{d}\theta\wedge\mathrm{d}r| \tag{33}$$

for Minkowski plane.

The Holmes-Thompson area $\text{HT}^2(U)$ of a measurable set U in a Minkowski plane is defined as $\text{HT}^2(U) := \frac{1}{\pi}\int_{D^*U} |\omega_0|^2$, where ω_0 is the natural symplectic form on the cotangent bundle of $\mathbb{R}^2$ and $D^*U := \{(x,\xi)\in T^*\mathbb{R}^2 : F^*(\xi)\leq 1\}$. To study it from the perspective of integral geometry, we need to introduce a symplectic form ω to the space of affine lines $\overline{Gr_1(\mathbb{R}^2)}$ and construct an invariant measure based on ω.

7. Application to HT area and related

Now let us see the Crofton formula for Minkowski plane, which is

$$\text{Length}(\gamma) = \frac{1}{4}\int_{\overline{Gr_1(\mathbb{R}^2)}} \#(\gamma\cap l)|\omega|. \tag{34}$$

To prove this, it is sufficient to show that it holds for any straight line segment

$$L : [0,||p_2-p_2||] \to \mathbb{R}^2, L(t) = p_1 + \frac{p_2-p_1}{||p_2-p_1||}t, \tag{35}$$

starting at p_1 and ending at p_2 in $\mathbb{R}^2$. First, using the diffeomorphism between the circle bundle and co-circle bundle, which is

$$\begin{aligned} \varphi_F &: S\mathbb{R}^2 \to S^*\mathbb{R}^2 \\ \varphi_F(x,\xi) &= (x, \mathrm{d}F_\xi), \end{aligned} \tag{36}$$

we can obtain a fact that

$$\begin{aligned} \int_{L\times\{\frac{p_2-p_1}{||p_2-p_1||}\}} \varphi_F^*\alpha_0 &= \int_{\varphi_F(L\times\{\frac{p_2-p_1}{||p_2-p_1||}\})} \alpha_0 \\ &= \int_0^{||p_2-p_1||} \alpha_{0\mathrm{d}F_{\frac{p_2-p_1}{||p_2-p_1||}}}\left(\left(\frac{p_2-p_1}{||p_2-p_1||}, 0\right)\right)\mathrm{d}t \\ &= \int_0^{||p_2-p_1||} \mathrm{d}F_{\frac{p_2-p_1}{||p_2-p_1||}}\left(\frac{p_2-p_1}{||p_2-p_1||}\right)\mathrm{d}t, \end{aligned} \tag{37}$$

where α_0 is the tautological one-form, precisely $\alpha_{0\xi}(X) := \xi(\pi_{0*}X)$ for any $X \in T_\xi T^*\mathbb{R}^2$, and $\mathrm{d}\alpha_0 = \omega_0$. Applying the basic equality that $\mathrm{d}F_\xi(\xi) = 1$, which is derived from the positive homogeneity of F, for all $\xi \in S\mathbb{R}^2$, the above quantity becomes $\int_0^{||p_2-p_1||} 1\mathrm{d}t$, which equals $||p_2-p_1||$.

Let $R := \{\xi_x \in S^*\mathbb{R}^2 : x \in \overline{p_1 p_2}\}$ and $T = \{l \in \overline{Gr_1(\mathbb{R}^2)} : l \cap \overline{p_1 p_2} \neq \emptyset\}$, and p' is the projection (composition) from $S^*\mathbb{R}^2$ to $\overline{Gr_1(\mathbb{R}^2)}$.

Apply the above fact and $p'^*\omega = \omega_0$,

$$\begin{aligned} \int_T |\omega| = \int_{p'(R)} |\omega| = \int_R |p'^*\omega| &= \int_R |\omega_0| \\ &= \left|\int_{R^+} \omega_0\right| + \left|\int_{R^-} \omega_0\right| \\ &= \left|\int_{\partial R^+} \alpha_0\right| + \left|\int_{\partial R^-} \alpha_0\right| \\ &= 4||p_2-p_1||. \end{aligned} \tag{38}$$

Thus, we have shown the Crofton formula for Minkowski plane.

Furthermore, combining with (33), we have

$$\frac{1}{4}\int_{l\in\overline{Gr_1(\mathbb{R}^2)}} \#(\gamma\cap l)|\Omega| = \frac{1}{4}\int_{\overline{Gr_1(\mathbb{R}^2)}} \#(\gamma\cap l)|\omega|, \tag{39}$$

where $\Omega = g(\theta)\mathrm{d}\theta\wedge\mathrm{d}r$. Then, by the injectivity of cosine transform in Ref. [24], $|\Omega| = |\omega|$.

To obtain the HT area, one can define a map

$$\begin{aligned} \pi &: \overline{Gr_1(\mathbb{R}^2)} \times \overline{Gr_1(\mathbb{R}^2)} \backslash \Delta \to \mathbb{R}^2 \\ \pi(l,l') &= l \cap l', \end{aligned} \tag{40}$$

extended from Alvarez's construction of taking intersections. The following theorem can be obtained.

Theorem 7.1. *For any bounded measurable subset U of a Minkowski plane, we have*

$$\mathrm{HT}^2(U) = \frac{1}{2\pi}\int_{x\in\mathbb{R}^2} \chi(x \cap U)|\pi_*\Omega^2|. \tag{41}$$

Proof. On the one hand,

$$\frac{1}{\pi}\int_{D^*U} \omega_0^2 = \frac{1}{\pi}\int_{\partial D^*U} \omega_0^2 = \frac{1}{\pi}\int_{S^*U} \alpha_0 \wedge \omega_0. \tag{42}$$

On the other hand,

$$\begin{aligned}
\tfrac{1}{\pi}\int_{x\in\mathbb{R}^2} \chi(x\cap U)\pi_*\Omega^2 &= \tfrac{1}{\pi}\int_{\left\{(l,l')\in \overline{Gr_1(\mathbb{R}^2)}\times\overline{Gr_1(\mathbb{R}^2)}\setminus\Delta : l\cap l'\in U\right\}} \Omega^2 \\
&= \tfrac{1}{\pi}\int_{\{(l,l')\in \overline{Gr_1(\mathbb{R}^2)}\times\overline{Gr_1(\mathbb{R}^2)}\setminus\Delta : l\cap l'\in U\}} \omega^2 \\
&= \tfrac{1}{\pi}\int_{\mathbb{T}^*U\setminus\{(x,\xi,\xi):\xi\in S_x^*U\}} p'^*\omega^2 \\
&= \tfrac{1}{\pi}\int_{\mathbb{T}^*U\setminus\{(x,\xi,\xi):\xi\in S_x^*U\}} \omega_0^2 \\
&= \tfrac{2}{\pi}\int_{\{(x,\xi,\xi):\xi\in S_x^*U\}} \alpha_0\wedge\omega_0 \\
&= \tfrac{2}{\pi}\int_{S^*U} \alpha_0\wedge\omega_0,
\end{aligned} \tag{43}$$

where

$$\mathbb{T}^*U := \left\{(x,\xi,\xi') : \xi,\xi' \in S_x^*U\right\}. \tag{44}$$

So the claim follows.

Remark 7.2. Lagrangian structure provides the underlying differential structure needed to measure the Holme-Thompson area in integral geometry and therefore is essential and doundamental in integral geometry. For Finsler manifolds, real or complex, it is necessary to analyze the Lagrangian structure of the Finsler manifolds, in the forms of symplectic structure and Kahler structure, and many Finsler manifolds may not have a Lagrangian structure, about which one can refer to Ref. [25]. However, for smooth projective Finsler spaces, the integral geometry formulas have been studied in Ref. [26], for instance.

Acknowledgements

The author would like to thank his family for their constant support for his academic career since his doctoral study in the USA, for the partial support by the National Science Foundation, and for the partial support by Air Force Office of Scientific Research under Grant AFOSR 9550-12-1-0455, and the author would also like to give thanks to Dr. P. Dang. Besides, the author would like to thank the reviewer for his or her helpful comments.

Author details

Yang Liu[1,2]

Address all correspondence to: yliu@msu.edu

1 Department of Mathematics, Michigan State University, East Lansing, MI, USA

2 School of Mathematics, Sun Yat-sen University, Guangdong, P.R. China

References

[1] Elie Cartan. Les espaces de finsler. *Bulletin of the American Mathematical Society*. 40:521–522, 1934. DOI: 10.1090/S0002-9904-1934-05891-9 PII.

[2] Hanno Rund. *The Differential Geometry of Finsler Spaces*. Springer, 1959.

[3] Zhong Chunping and Zhong Tongde. Horizontal laplace operator in real finsler vector bundles. *Acta Mathematica Scientia*, 28(1):128–140, 2008.

[4] Shiing-Shen Chern and Zhongmin Shen. *Riemann-Finsler Geometry*. World Scientific, 2005.

[5] Gheorghe Munteanu. Complex finsler spaces. In *Complex Spaces in Finsler, Lagrange and Hamilton Geometries*, pages 55–90. Springer, 2004.

[6] Tadashi Aikou. On complex finsler manifolds. *Reports of Kagoshima University*, 24:9–25, 1991.

[7] T Kawaguchi. On the application of finsler geometry to engineering dynamical systems. *Periodica Mathematica Hungarica*, 8(3–4):281–289, 1977.

[8] Yang Liu. On the lagrangian subspaces of complex minkowski space. *Journal of Mathematical Sciences: Advances and Applications*, 7(2):87–93, 2011.

[9] Erwin Kreyszig. *Introductory Functional Analysis with Applications*, volume 81. Wiley New York, 1989.

[10] Eberhard Zeidler. *Nonlinear Functional Analysis and Its Applications*, volume 2. Springer Science & Business Media, 1989.

[11] Herbert Weisberg. *Central Tendency and Variability*. Number 83. Sage, 1992.

[12] Yang Liu. On explicit holmes-thompson area formula in integral geometry. *International Mathematical Forum*, 2011. Accepted for publication. arXiv:1009.5057.

[13] Patrick L Combettes and Valérie R Wajs. Signal recovery by proximal forward-backward splitting. *Multiscale Modeling & Simulation*, 4(4):1168–1200, 2005.

[14] JD Clayton. On finsler geometry and applications in mechanics: Review and new perspectives. *Advances in Mathematical Physics*, 2015, 2015.

[15] Yu Xia. Newton's method for the ellipsoidal l_p norm facility location problem. In *Computational Science–ICCS 2006*, volume 3991, pages 8–15, 2006.

[16] Peter L Antonelli and Radu Miron. *Lagrange and Finsler Geometry: Applications to Physics and Biology*, volume 76. Springer Science & Business Media, 2013.

[17] Simon Foucart and Ming-Jun Lai. Sparsest solutions of underdetermined linear systems via q-minimization for 0< q<=1. *Applied and Computational Harmonic Analysis*, 26(3):395–407, 2009.

[18] Ming-Jun Lai and Yang Liu. The null space property for sparse recovery from multiple measurement vectors. *Applied and Computational Harmonic Analysis*, 30(3):402–406, 2011.

[19] Ming-Jun Lai and Yang Liu. The probabilistic estimates on the largest and smallest q-singular values of random matrices. *Mathematics of Computation*, 2014. doi:10.1090/S0025-5718-2014-02895-0

[20] David Dai-Wai Bao. *A Sampler of Riemann-Finsler Geometry*, volume 50. Cambridge University Press, 2004.

[21] Luis A Santaló. Integral geometry in hermitian spaces. *American Journal of Mathematics*, 74(2):423–434, April, 1952.

[22] Andreas Bernig and Joseph HG Fu. Hermitian integral geometry. *Annals of Mathematics*, 173(2):907–945, 2011.

[23] Yang Liu. On the range of cosine transform of distributions for torus-invariant complex minkowski spaces. *Far East Journal of Mathematical Sciences*, 39(2):733–753, 2010.

[24] Helmut Groemer. *Geometric Applications of Fourier Series and Spherical Harmonics*, volume 61. Cambridge University Press, 1996.

[25] Yang Liu. On the kähler form of complex l^{p} space and its lagrangian subspaces. *Journal of Pseudo-Differential Operators and Applications*, 6(2):265–277, 2015.

[26] Israel M Gelfand, James Lepowsky, and Mikhail Smirnov. *The Gelfand Mathematical Seminars, 1993–1995*. Springer Science & Business Media, 2012.

[27] Yang Liu. Probabilistic estimates of the largest strictly convex singular values of pregaussian random matrices. *Journal of Mathematics and Statistics*, 11(1):7–15, 2015.

Chapter 3

Topology and Integrability in Lagrangian Mechanics

Leo T. Butler

Additional information is available at the end of the chapter

http://dx.doi.org/10.5772/66147

Abstract

This chapter reviews complete integrability in the setting of Lagrangian/Hamiltonian mechanics. It includes the construction of angle-action variables in illustrative examples, along with a proof of the Liouville-Arnol'd theorem. Results on the topology of the configuration space of a mechanical (or Tonelli) Hamiltonian are reviewed and several open problems are high-lighted.

Mathematics Subject Classication (2010): 37J30; 53C17, 53C30, 53D25

Keywords: Hamiltonian mechanics, Lagrangian mechanics, integrability, topological obstructions, topological entropy

1. Introduction

Lagrangian mechanics employs the least-action principle to derive Newton's equations from a scalar function, the action function L. In classical mechanics, L is the difference of kinetic and potential energies and therefore appears as an artifice. It is somewhat mysterious, then, that the reformulation of Newtonian mechanics in terms of momentum and position, rather than velocity and position as in Lagrangian mechanics, leads immediately to the total energy function H and a plethora of geometric structure that is hidden in the native setting.

Due to the advantages of the Hamiltonian perspective, this chapter studies Lagrangian systems from this dual point of view. The organization of the chapter is this: Section 2 recalls the classic construction of angle-action variables in 1 degree of freedom via several examples, then states and proves the Liouville- Arnol'd theorem; Section 3 discusses the relationship between the topology of the configuration space and the existence of integrable mechanical systems; and it reviews several constructions of integrable systems whose configuration space is the sphere or torus. Section 3 provides a number of open problems that may stimulate interested researchers or students.

2. Integrability in Hamiltonian mechanics

2.1. Integrability in 1 degree of freedom

One of the central problems in classical mechanics is the *integrability* of the equations of motion. The classical notion of integrability is loosely related to exact solvability, and roughly corresponds to the ability to solve a system of differential equations by means of a finite number of integration steps.

2.1a. *Example: Harmonic oscillator* Let us take the simple harmonic oscillator, or an idealized Hookean spring-mass system, with mass m and spring constant k. If q is the displacement from equilibrium and p the momentum, then the total energy is

$$H = \frac{1}{2m}p^2 + \frac{k}{2}q^2, \quad \text{and equations of motion are} \quad \left\{ \begin{array}{ccc} \dot{q} & = & p/m, \\ \dot{p} & = & -kq \end{array} \right\}. \tag{1}$$

The change of variables $(q, p) = (Q/\lambda, \lambda P)$ transforms the system to, with $\lambda = \sqrt[4]{km}$,

$$H = \frac{\omega}{2}(P^2 + Q^2), \quad \text{and equations of motion are} \quad \left\{ \begin{array}{ccc} \dot{Q} & = & \omega P, \\ \dot{P} & = & -\omega Q \end{array} \right\}, \tag{2}$$

where $\omega = \sqrt{k/m}$. A second change of variables $(Q, P) = (\sqrt{2I}\cos\theta, \sqrt{2I}\sin\theta,)$ transforms the system to

$$H = \omega I, \quad \text{and equations of motion are} \quad \left\{ \begin{array}{ccc} \dot{\theta} & = & \omega, \\ \dot{I} & = & 0 \end{array} \right\}. \tag{3}$$

The differential equations in (3) are trivial to integrate since the right-hand sides are constants. Let us explain the sequence of transformations. The change of coordinates $(q, p) \to (Q, P)$ is an area-preserving linear transformation that transforms the elliptical level sets of H into circles. The transformation $(Q, P) \to (\theta, I)$ is analogous to the introduction of polar coordinates–indeed the transformation $(r,\theta) = (\sqrt{2I},\theta)$ is a transformation to polar coordinates. Because the area form $\mathrm{d}P\,\mathrm{d}Q = r\ \mathrm{d}r\ \mathrm{d}\theta$, we see that the transformation $\mathrm{d}P\,\mathrm{d}Q = \mathrm{d}I = \mathrm{d}\theta$.

Therefore, the change of coordinates $(q, p) \to (\theta, I)$ not only reveals the exact solutions of the harmonic oscillator equations, it is area preserving.

Suppose that for some reason one did not know to introduce "polar" coordinates. One might still determine the change of coordinates using only that the transformation $(Q, P) \to (\theta, I)$ preserves area. Indeed, since $\mathrm{d}\ (P\,\mathrm{d}Q - I\,\mathrm{d}\theta) = 0$, there is a function $v = v(Q, \theta)$ such that $P\,\mathrm{d}Q - I\ \mathrm{d}\theta = dv$ or $P = \frac{\partial v}{\partial Q}$ and $I = -\frac{\partial v}{\partial \theta}$. Then, upon substituting the identity $P = v_Q$ into (2), one obtains

$$v \equiv \int_0^Q \sqrt{2H/\omega - Q^2}\,\mathrm{d}Q = \frac{1}{2}Q\sqrt{2H/\omega - Q^2} - (H/\omega)\ \arccos(Q/\sqrt{2H/\omega}), \tag{4}$$

where ≡ indicates that v equals the right-hand side up to the addition of a 2π-periodic function of θ.

If (Q,P) make a complete circuit around the contour $\{H = c\}$ then one obtains from (4) and the identity that $P = v_Q$ that

$$\Delta v = \oint_{\{H=c\}} P\,\mathrm{d}Q = (H/\omega)\,2\pi. \tag{5}$$

On the other hand, since $\mathrm{d}^2 = 0$ and I is held constant on the contour, Green's theorem implies that

$$\Delta v = \oint_{\{H=c\}} \mathrm{d}v + I\,\mathrm{d}\theta = \oint_{\{H=c\}} I\,\mathrm{d}\theta = 2\pi I. \tag{6}$$

Equating (5) and (6) shows that $H = \omega I$.

These calculations show that one may determine H as a function of I without explicit knowledge of the coordinate transformation $(Q,P) \to (\theta,I)$–but one does need to solve the Hamilton-Jacobi equation

$$H(Q, v_Q) = c, \tag{7}$$

for v, as performed in Eq. (4). At this point, if one wants to derive the change of coordinates from v, Eq. (4) shows that it is easier to write $v = v(Q, I)$, in which case $P\ \mathrm{d}Q + \theta\ \mathrm{d}I = \mathrm{d}v$ or

$$\theta = v_I = -\arccos\,(Q/\sqrt{2I}), \tag{8}$$

so $Q = \sqrt{2I}\cos(\theta)$ and $P = v_Q = \sqrt{2I}\sin(\theta)$.

Let it be observed that if, in Eq. (4), one had chosen the anti-derivative to be arcsin rather than –arccos, then Q would be $\sqrt{2I}\sin(\theta)$ and P would be $\pm\sqrt{2I}\cos(\theta)$. However, because $\mathrm{d}P\ \mathrm{d}Q = \mathrm{d}I\ \mathrm{d}\theta$, one would be obligated to choose the negative square root to define P; otherwise, $\mathrm{d}P\ \mathrm{d}Q = -\mathrm{d}I\ \mathrm{d}\theta$.

2.1b. *Example: the planar pendulum*. Let us take the idealized planar pendulum with a mass-less rigid rod of length l suspended at a fixed end with a bob of mass m at the opposite end (**Figure 1**). The total energy is

$$H = \frac{1}{2m}p^2 + mlg(1-\cos q), \quad \text{and} \quad \left\{\begin{matrix} \dot{q} & = & p/m, \\ \dot{p} & = & -mlg\sin q \end{matrix}\right\}. \tag{9}$$

To simplify the exposition, assume that the mass $m = 1$ and let $\omega^2 = 16lg$, where ω is 4 times the frequency of the linearized oscillations at $q = p = 0$. The substitution $q = 2Q$, $p = P/2$ transforms the Hamiltonian to

$$8H = P^2 + \omega^2\sin^2 Q, \quad \text{and} \quad \left\{\begin{matrix} \dot{Q} & = & P/4 \\ \dot{P} & = & -\omega^2\sin(2Q)/8 \end{matrix}\right\}. \tag{10}$$

If one tries to solve for a generating function $v = v(Q,I)$ of a coordinate change $(Q,P) \to (\theta,I)$ such that $H = H(I)$, then one obtains from $P = v_Q$ that

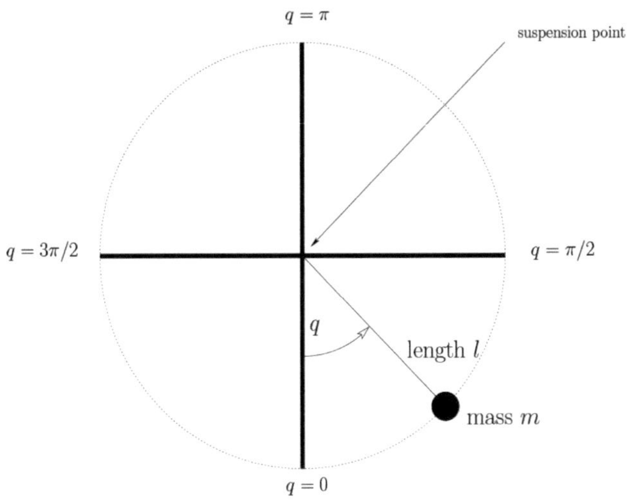

Figure 1. The planar pendulum with potential energy $V = mlg(1-\cos q)$.

$$\nu \equiv \int_0^Q \sqrt{8H-\omega^2 \sin Q}\mathrm{d}Q = \frac{\omega}{k} E(Q,k) \tag{11}$$

where ≡ indicates equality up to a 2π-periodic function of θ, $H = \omega^2/(8k^2)$ and E is the elliptic integral of the second kind defined by $E(x,\ k) = \int_0^x \sqrt{1-k^2 \sin^2 x}\ \mathrm{d}x$.

If (Q,P) make a complete circuit around the contour $\{H = c\}$, then one obtains from Eq. (11) that

$$\Delta\nu = 4\frac{\omega}{k} K(k) \tag{12}$$

where $K(k) = E(Q_+(k),k)$ and $Q_+(k) = \arcsin(1/k)$ if $k > 1$ and $\pi/2$ if $k < 1$ (in which case, K is the complete elliptic integral of the second kind). The area of the shaded region K in **Figure 2** shows the geometric meaning of $K(k)$ for $k > 1$. Along with the identity (6), one obtains

$$I = \frac{2}{\pi}\frac{\omega}{k} K(k), \tag{13}$$

which determines $H = H(I)$ implicitly.

Figure 3 graphs H as a function of I using the definition of I in (13) with $\omega = 1$, along with the graph for the harmonic oscillator. Although H appears to be a smooth function of I on the interval depicted, this is a numerical artifact. Indeed, there are two distinct proofs that H cannot be differentiable in I over the interval [0,1]. Without loss of generality, it is assumed that $\omega = 1$.

http://dx.doi.org/10.5772/66147

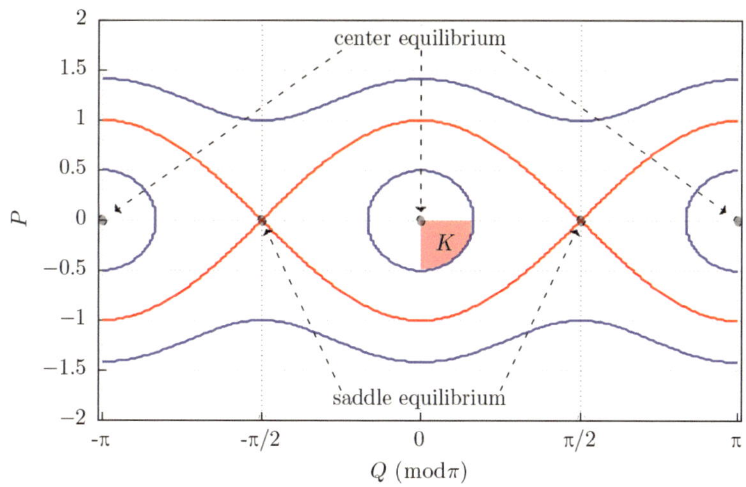

Figure 2. The contours of the pendulum Hamiltonian with $\omega = 1$ (9).

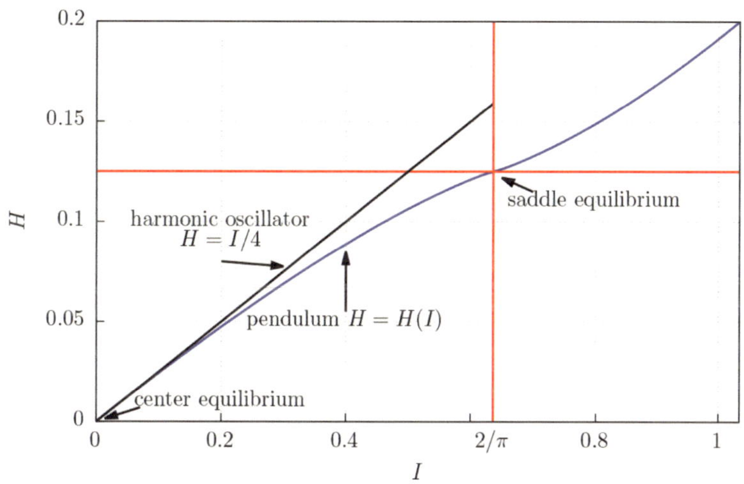

Figure 3. The graph of $H = H(I)$for the pendulum.

The first, calculus-based, proof is this: as $k \to 1^+ (H \to 1/8^-), \partial I/\partial k \to \infty$. If H is a differentiable function of I, then $\partial H/\partial I = 0$ at $I = 2/\pi (H = 1/8)$. But then the entire level set consists of fixed points, which is false.

The second, topological, proof is this: each level set $\{H = c\}, c < 1/8$, is connected; each level set for $c > 1/8$ has exactly two connected components (*c.f.* **Figure 2**). If the generating function v were differentiable in (Q, I) on any rectangle containing $\mathbf{R}/\pi\mathbf{Z} \times \{2/\pi\}$, then Eq. (13) would determine a homeomorphism $H = H(I)_,$ and so the level sets of H would remain connected on either side of the critical level at height 1/8. Absurd.

To derive the change of coordinates $(Q, P) \to (\theta, I)$ from the generating function ν, one uses the identity $\theta = \nu_I$ and properties of the elliptic integrals to deduce

$$\theta = \frac{\pi}{2}\frac{F(Q,\, k)}{(Q_+(k), k)} \quad \Rightarrow \quad Q = \mathrm{am}_k\left(\frac{2F_+}{\pi}\theta\right) \tag{14}$$

where $F(x, k) = \int_0^x \mathrm{d}x/\sqrt{1 - k^2 \sin^2 x}$ is the elliptic integral of the first kind, $F_+ = F(Q_+(k), k)$ and $\mathrm{am}_k(u)$ is the Jacobian amplitude function, a local inverse to F ([1], Chapter 2). Along with $P = \nu_Q$, (14) implies that

$$P = \frac{\omega}{k}\mathrm{dn}_k\left(\frac{2F_+}{\pi}\theta\right), \tag{15}$$

where $\mathrm{dn}_k(u)$is the Jacobian elliptic function.

2.1c. *Example: a mechanical system*. Let $V = V(Q)$ be a smooth potential function of a 1-degree-of-freedom Hamiltonian system with

$$H = \frac{1}{2}P^2 + V(Q). \tag{16}$$

If one attempts to find the generating function $\nu = \nu(Q, I)$ of an area-preserving transformation $(Q, P) \to (\theta, I)$ that transforms $H = H(I)$, then one deduces that

$$\nu \equiv \int_{Q_0}^{Q} \sqrt{2(H - V(Q))}\, \mathrm{d}Q, \tag{17}$$

up to a function depending only on I. Then, in a complete circuit around the connected contour $\{H = c\}$, one has $2\pi = \Delta\theta = \Delta\nu_I$ identically, so

$$2\pi I = \oint_{\{H=c\}} P\, \mathrm{d}Q. \tag{18}$$

and, upon solving (18) for $H = H(I)$, one inverts

$$\theta = \frac{1}{\sqrt{2}}\int_{Q_0}^{Q} \frac{H_I}{\sqrt{H - V(Q)}}\mathrm{d}Q, \tag{19}$$

to obtain $Q = Q(\theta, I)$, and finally $P = \nu_Q$ yields $P = P(\theta, I)$. Since the change of coordinates is area-preserving, the Hamiltonian form of the equations of motion are preserved, so the resulting equations are

$$H = H(I) \quad \text{and} \quad \left\{\begin{matrix} \dot{\theta} & = & \partial H/\partial I \\ \dot{I} & = & 0 \end{matrix}\right\}. \tag{20}$$

2.2. The generating function

The above three examples use a generating function $\nu = \nu(Q, I)$ of a mixed system of coordinates in order to create an area-preserving change of coordinates to angle-action variables $(\theta,\, I)$.

http://dx.doi.org/10.5772/66147

2.2a. *Question: why do the angle-action variables exist?* In order to understand the generating function, it is necessary to clarify the existence of the coordinates (θ, I), which are commonly called angle-action variables. Let $H : X \to \mathbf{R}$ be a smooth function from an oriented surface X to the reals. If it is assumed that $A \subset X$ is an open, connected, saturated $(H^{-1}(H(A)) \cap A = A)$ subset of the domain of $H, H|A$has no critical points and $H|A$ is proper, then $H|A$ is a submersion onto the interval $B = H(A) \subset \mathbf{R}$. Since $H|A$ is proper, for each $b \in B$, the level set $(H|A)^{-1}(b)$ is a compact one-manifold and hence its components are circles. Since A is connected and $H|A$ is critical-point free, the level set must be connected, so it is a circle. Therefore, the submersion theorem implies that A is diffeomorphic to $A' = S^1 \times B$.[1] To make this system of coordinates concrete, note that there is a complete vector field U on A such that $\mathrm{d}H(U) \equiv 1$. Let $\gamma \subset A$ be a segment of an integral curve of U which is maximal (i.e. an integral curve that strictly contains γ intersects $X - A$). For each $a \in A$, let $t = t(a)$ be the time along the flow line of the Hamiltonian vector field X_H beginning at the initial condition $\gamma \cap H^{-1}(H(a))$. The function t is multi-valued, since the flow line is closed, so it should be considered as a function on the universal cover of A.

Since the tangent space at $a \in A$ is spanned by X_H and U, Ω is determined by $\Omega(X_H, U)$. But $Q(X_H, U) = -\mathrm{d}H(U) = -1$, so $\Omega = \mathrm{d}H\mathrm{d}t$.

Let T be such that $2\pi T$ is the least period of the function t (i.e. the first return time to γ). Then $T = T(H)$ is a function of H alone. Define θ by

$$\theta = t/T(H) \pmod{2\pi} \quad \text{and } I \text{ by} \quad \mathrm{d}I = T(H)\,\mathrm{d}H. \tag{21}$$

The function θ is the normalized time along the flow lines of Hamiltonian vector field X_H, while $\mathrm{d}H/\mathrm{d}I = 1/T(I)$ is the frequency. One computes that the oriented area form $\Omega = \mathrm{d}H dt = \mathrm{d}I\mathrm{d}\theta$. Moreover, in the coordinates (θ, I), the Hamiltonian vector field

$$X_H = \left\{ \begin{array}{ccccc} \dot{\theta} & = & 1/T(H) & = & \dfrac{\mathrm{d}H}{\mathrm{d}I}, \\ \dot{I} & = & 0 & & \end{array} \right\}. \tag{22}$$

This proves the existence of an area-preserving diffeomorphism $\phi : D \times \mathbf{S}^1 \to A$, where $D \subset \mathbf{R}$ is an open interval, such that the Hamiltonian H is transformed to a function of I alone; and ϕ is as smooth as H and the area form Ω are (e.g. if both are real-analytic, then ϕ is real-analytic).

2.2b. Question: what kind of "function" is v? In the first instance, v is not single-valued. Indeed, one postulates the area-preserving change of coordinates $\phi : (Q, P) \to (\theta, I)$ to deduce that

$$\mathrm{d}(P\mathrm{d}Q + \theta\mathrm{d}I) = 0, \tag{23}$$

so that *locally* there is a function v such that

$$P\,\mathrm{d}Q + \theta\mathrm{d}I = \mathrm{d}v. \tag{24}$$

[1] If one prefers a purely "elementary" proof, one might apply the inverse function theorem at this point.

But since θ is an angle variable, this equation can only hold globally modulo $2\pi\mathbf{Z}\, dI$. So, in this formulation of the generating function, v can only be defined globally modulo $2\pi\mathbf{Z}\, I$. Or, equivalently, v_I is a function with values in the circle $\mathbf{R}/2\pi\mathbf{Z}$.

The way to resolve these ambiguities or difficulties is simple: the domain of the change of coordinates ϕ must be non-simply connected (a disjoint union of open annuli, in fact, as can be deduced from the discussion above) and so one should view (24) as holding globally on the universal cover of this annulus where θ is a single-valued real function (*c.f.* 21). In this case, the lift of a closed contour $\{H = c\}$ is a path that projects to the contour and whose endpoints differ by a deck transformation–which in the angle-action coordinates is $(\theta, I) \to (\theta + 2\pi, I)$. Since I is constant along this path, the path integral of PdQ equals the path integral of dv, i.e. Δv, the change in v from one preimage to its translate. With this understanding, Eq. (18) is correct. And, indeed, one sees that the integral in Eq. (17) is defined not on the domain of the coordinate change ϕ but on its universal cover; the same is true for the integral in Eq. (19), but the marvellous fact about that integral is that it is 2π-periodic: this follows from the observation that $\Delta\theta = 2\pi$ identically around a closed connected contour in $\{H = c\}$.

So to answer the question that started the section, the generating function v is a function defined on the universal cover of the union of regular compact levels of H which implicitly defines a 2π-periodic change of coordinates to "angle-action" variables (θ, I).

2.3. Integrability in 2 or more degrees of freedom and Tonelli Hamiltonians

Integrability in 2 or more degrees of freedom is substantially more involved than the case of 1 degrees of freedom. Of course, a sum of n distinct, non-interacting 1-degree-of-freedom Hamiltonians is a simple case; and upon reflection, a not-so- simple case, because this condition is not coordinate independent. Indeed, a necessary and sufficient condition is that the Hamiltonian vector field be Hamiltonian with respect to two distinct non-degenerate Poisson brackets $\{,\}_i$ that are compatible in the sense that the linear space spanned by the brackets is a space of Poisson brackets, and maximal in the sense that a "recursion" operator naturally defined from the two brackets has a maximal number of functionally independent eigenvalue fields [2].

Let us turn now to a definition which generalizes mechanical Hamiltonians.

Definition 2.1 *(Tonelli Hamiltonian). Let Σ be a smooth n-manifold and $T^*\Sigma$ its cotangent bundle. A smooth function $H : T^*\Sigma \to \mathbf{R}$ which satisfies* $(\mathrm{T}_1) H|T^*_x\Sigma$ *is strictly convex for each* $x \in \Sigma$*; and* (T_2) $H(x, tp)/t \to \infty$ *uniformly as* $t \to \infty$*, is called a Tonelli Hamiltonian.*

As noted, Tonelli Hamiltonians are natural generalizations of mechanical systems. For this reason, Σ will be referred as the configuration space of the Hamiltonian H.

If Q_i are coordinates on Σ and $\Theta = \sum_i P_i\, dQ_i$ are the coordinates of the 1-form Θ, then the canonical symplectic structure $\Omega = d\Theta = \sum dP_i \wedge dQ_i$ on $T^*\Sigma$. The symplectic form Ω equips the space of smooth functions on $T^*\Sigma$ with a Poisson bracket denoted $\{,\}$ that satisfies

$$\{P_i, Q_j\} = -\{Q_j, P_i\} = \delta_{ij} \quad \{Q_i, Q_j\} = \{P_i, P_i\} = 0 \tag{25}$$

for all i, j. The Poisson bracket is fundamental to Hamiltonian mechanics. For each smooth function H, one has a smooth vector field $X_H = \{H, \}$, and the skew symmetry of the bracket implies that H is preserved by the flow. One says that H_1 and H_2 Poisson commute if $\{H_1, H_2\} \equiv 0$.

A fundamental result in Hamiltonian mechanics is the Liouville-Arnol'd theorem, which provides a semi-local description of a completely integrable Hamiltonian and the Poisson bracket.

Theorem 2.1 *(Liouville-Arnol'd). Let $H : T^*\Sigma \to \mathbf{R}$ be a smooth Hamiltonian. Assume there exists n functionally independent, Poisson commuting conserved quantities $F = (F_1 = H, \ldots, F_n) : T^*\Sigma \to \mathbf{R}^n$.* If $L \subset F^{-1}(c)$ *is a compact component of a regular level set, then there is a neighbourhood W* of L *and a diffeomorphism $\phi = (\theta, I) : \mathbf{T}^n \times \mathbf{B}^n \to W$ such that*

$$F = F(I) \quad \{I_i, \theta_j\} = \delta_{ij}, \quad \{I_i, I_j\} = \{\theta_i, \theta_j\} = 0,$$
$$X_{F_i} = \sum \frac{\partial F_i(I)}{\partial I_j} \frac{\partial}{\partial \theta_j},$$

that maps L to $\mathbf{T}^n \times \{0\}$.

In such a situation, it is said that $H$is Liouville, or completely, integrable. The torus $\mathbf{T}^n \times \{I_0\}$ is a *Liouville torus*, the neighbourhood $\mathbf{T}^n \times \mathbf{B}^n$ is a *toroidal ball* and the conserved quantities are *first integrals*. Systems with k first integrals, of which $l < k$ Poisson commute with all k first integrals, where $k + l = 2n$ are called *non-commutatively integrable*; when $k = 2n-1$, the system is also called *super-integrable c.f.* [3, 4].

There are several proofs of the Liouville-Arnol'd theorem in the literature. The basic ideas are already captured in the one-dimensional case discussed in Section 2.2.

It can be assumed, without loss, that $L = F^{-1}(c)$. Since $c \in \mathbf{R}^n$ is a regular value of F, the submersion theorem implies that there is an open neighbourhood C of c consisting of regular values of F and the open set $F^{-1}(C)$ is diffeomorphic to $L \times C$. Therefore, there is a smooth n-dimensional submanifold $M \subset F^{-1}(C)$ such that M transversely intersects each level set $L_f = F^{-1}(f), f \in C$. Possibly by shrinking the open set C, it can be assumed that M is Lagrangian: $\Omega|M \equiv 0$.[2]

Because the functions $F_i, \ldots, F_n$ Poisson commute and are functionally independent, the Hamiltonian vector fields $X_{F_1}, \ldots, X_{F_n}$ span the tangent space $T_x L_f$, for each $x \in L_f, f \in C$. Because L_f is compact, each vector field is complete, so there is a well-defined flow map $\phi_{F_i} : \mathbf{R} \times F^{-1}(C) \to F^{-1}(C)$. Because $F_i, \ldots, F_n$ Poisson commute, the respective flow maps commute, so there is an action of $\mathbf{R}^n$ on $F^{-1}(C)$ defined by

$$\phi^t = \phi_{F_n}^{t_n} \circ \cdots \circ \phi_{F_1}^{t_1} \tag{26}$$

for all $t \in \mathbf{R}^n$. Define a map

[2]The existence of M is a consequence of Darboux's theorem. Of course, a less elementary proof would appeal to Weinstein's theorem and Moser's isotopy lemma.

$$\Phi(t,\ m) = \phi^t(m), \quad t \in \mathbf{R}^n, m \in M. \tag{27}$$

This is a smooth map which is a local diffeomorphism of $\mathbf{R}^n \times M$ with $F^{-1}(C)$. Indeed, ϕ^t carries each level L_f into itself and carries M into a submanifold $\phi^t(M)$ transverse to L_f at $\phi^t(m)$; on the other hand, the derivative of ϕ^t with respect to t is a surjective linear map onto $T_{\phi^t(m)}L_f$. Therefore, $d\Phi$ is surjective, so injective, hence Φ is a local diffeomorphism onto its image. Compactness and connectedness of the levels L_f imply that the image of Φ is $F^{-1}(C)$.

For each $m \in M$, let $P(m) \subset \mathbf{R}^n$ be the set of t such that $\Phi(t,\ m) = m$. Since each level set is compact, $P(m)$ is a discrete subgroup of $\mathbf{R}^n$ isomorphic to $\mathbf{Z}^n$. This is the "period lattice" of the action ϕ. If one selects a basis of $P(m)$, one obtains a map $M \to \mathrm{GL}(n; \mathbf{R}), m \to 2\pi T(m)$. The implicit function theorem implies that there is a smooth map amongst these maps. Moreover, since $F|M$ is a bijection onto its image, one can take the components of F as coordinates on M, or in other words, $T = T(F)$.

Define functions $\theta = (\theta_1, ..., \theta_n)$ by

$$\theta = T(F)^{-1} \cdot t \pmod{2\pi}, \quad \theta : \mathbf{R}^n \times C \to \mathbf{R}^n / 2\pi\mathbf{Z}^n. \tag{28}$$

The flow map Φ therefore induces a diffeomorphism $F^{-1}(C) \to \mathbf{T}^n \times C : x \to (\theta(x), F(x))$.

To complete the proof, one might show that each vector field $\partial/\partial\theta_i$ is Hamiltonian with Hamiltonian function I_i and that F is functionally dependent on I so that (θ, I) is a canonical system of coordinates on $F^{-1}(C)$. This is performed indirectly. Define the functions $I_i = I_i(F)$ by

$$2\pi I_i = \oint_{\Gamma_i(F)} \xi, \tag{29}$$

where $\xi = P \cdot dQ$ is the primitive of the symplectic form Ω and $\Gamma_i(F)$ is the cycle on L_F on which θ_i increases from 0 to 2π and the other angle variables are held equal to 0. To show that (θ, I) is a system of coordinates on $F^{-1}(C)$, one computes the Jacobian $[\partial I_i / \partial F_j]$:

$$2\pi \frac{\partial I_i}{\partial F_j} = \lim_{s \to 0} \frac{1}{s} \int\int_{C_j(F,s)} \Omega, \tag{30}$$

where, in the (t, F) coordinate system,

$$C_j(F,\ s) = \{(uT(F + ve_j)_i, F + ve_j) | u \in [0, 2\pi],\ v \in [0, s]\}$$

is the "cylinder" obtained by sweeping out the cycles $\Gamma_j(F + ve_j)$ as the j-th component of Fincreases from F_j to $F_j + s$, and T_i is the i-th column of the period matrix T. Since

$$\Omega\left(\frac{\partial}{\partial F_j}, \frac{\partial}{\partial t_k}\right) = \frac{\partial F_k}{\partial F_j} = \delta_{jk}, \tag{31}$$

which implies

$$\frac{\partial I_i}{\partial F_j} = T_{ji}. \quad (32)$$

Since the period matrix T is non-singular, the transformation $(\theta, F) \to (\theta, I)$ is a diffeomorphism.

Finally, the functions $I_1, \ldots, I_n$ Poisson commute and since M is Lagrangian, the functions $t_i, \ldots, t_n$ Poisson commute, which implies $\theta_1, \ldots, \theta_n$ Poisson commute. And, since $\{F_i, t_j\} = \delta_{ij}$, this implies that $\{I_i, \theta_j\} = \delta_{ij}$.

The remainder of the theorem follows from the fact that the angle-action coordinates (θ, I) are canonical and $F = F(I)$.

3. Topology of configuration spaces

The central problem in the theory of completely integrable Tonelli Hamiltonian systems is to

Problem 3.1. *Determine necessary conditions on the configuration space Σ for the existence of a completely integrable Tonelli Hamiltonian H.*

This is a broad, overarching problem which has motivated research by many authors over an almost 40-year period, including many of the author's publications. It is helpful to pose several sub-problems which address aspects of this problem and that appear to be amenable to solution. The remainder of this section is devoted to an elaboration of this problem, along with known results. We start with two-dimensional configuration spaces.

3.1. Surfaces of genus more than one

As a rule, completely integrable Tonelli Hamiltonians are quite rare, as are the configuration spaces Σ which support such Hamiltonians. Indeed, in two dimensions, the compact surfaces that are *known* to support a completely integrable Tonelli Hamiltonian are the 2-sphere, $\mathbf{S}^2$, the 2-torus $\mathbf{T}^2$ and their non-orientable counterparts. With some quite mild restrictions on the singular set–called condition $\aleph-$, and assuming that the Hamiltonian is Riemannian, Bialy has proven these are the only compact examples [5]. This extended an earlier result of V. V. Kozlov [6]; the author has obtained a similar result for super-integrable Tonelli Hamiltonians [7].

V. Bangert has suggested to the author that Bialy's argument should extend to prove the non-existence of a C^2 integral that is independent of the Hamiltonian when Σ is a compact surface of negative Euler characteristic (*c.f.* [8]). The idea of such a proof would be the following (assuming that H is Riemannian): Suppose that H enjoys a C^2 integral F that is independent on a dense set, hence that the union of Liouville tori is dense. Let $\Gamma \subset H^{-1}(\frac{1}{2})$ be the union of orbits which project to minimizing geodesics. It is known, due to results of Manning and Katok [9, 10], that $\overline{\Gamma}$ contains a hyperbolic invariant set Λ on which the flow is conjugate to a horseshoe. Let $\lambda \subset \Lambda$ be a closed orbit of the geodesic flow of period T. Since the union of Liouville tori is dense, for each $\epsilon > 0$, there is a Liouville torus $L_{\lambda,\epsilon}$ that contains an orbit of the geodesic flow that remains within a distance ϵ of λ over the interval $[0,T]$. Hence, $\pi_1(L_{\lambda,\epsilon})$ has a homotopy class mapping onto λ. Since λ is minimizing, it has no conjugate points and so for ϵ sufficiently small, the same is true for the orbit on $L_{\lambda,\epsilon}$ over the time interval $[0,T]$. This implies that the image of $\pi_1(L_{\lambda,\epsilon})$ is (free) cyclic and the kernel is generated by a cycle that bounds a disc–in classical terminology, this means that $L_{\lambda,\epsilon}$ is *compressible*. It follows that $L_{\lambda,\epsilon}$ bounds a solid torus $T_\lambda \cong \mathbf{T}^1 \times \mathbf{B}^2$ that is invariant for the geodesic flow. The integral $F|T_\lambda$ induces a singular fibration of the solid torus by invariant 2-tori.

Thus, for each closed orbit λ in the hyperbolic invariant set Λ, we have produced an invariant solid torus T_λ that shadows λ–at least in some rough, homotopic sense. This fact alone should suffice to achieve a contradiction.

Problem 3.2. *Let Σ be a compact surface of negative Euler characteristic. Extend the above argument to prove the non-existence of a smooth Tonelli Hamiltonian $H : T^*\Sigma \to \mathbf{R}$ with a second C^2 integral F that is independent on a dense set; or give an example of a completely integrable Tonelli Hamiltonian $H : T^*\Sigma \to \mathbf{R}$.*

V. Bangert proposes similar problems in his contribution in ([8], Problems 1.1, 1.2).

There is a similar, but possibly more accessible, problem for twist maps. Recall that if we discretize time, the notion of a Tonelli Hamiltonian is replaced by that of a *twist map* $f : T^*\Sigma \to T^*\Sigma$ *which is a symplectomorphism that satisfies a condition analogous to* T_1. If f enjoys n independent, Poisson commuting first integrals, then the Liouville-Arnol'd theorem implies that some power of f acts a translation on the Liouville tori. We noted above that the Hamiltonian flow of a Tonelli Hamiltonian has a horseshoe on an energy level.

Problem 3.3. *Let $f : T^*\mathbf{T}^1 \to T^*\mathbf{T}^1$ be a twist map. If f has a horseshoe and a C^1 first integral F, is F necessarily constant on an open set?*

3.2. The 2-torus

Let us turn now to the torus. The 2-torus $\mathbf{T}^2$ admits a family of completely integrable Riemannian Hamiltonians which are called *Liouville*. These are of the form

$$H = \frac{p_x^2 + p_y^2}{2[f(x) + g(y)]} \qquad F = \frac{g(y)p_x^2 - f(x)p_y^2}{f(x) + g(y)} \tag{33}$$

where $f, g : \mathbf{T}^1 \to \mathbf{R}$ are smooth positive functions and (x, y, p_x, p_y) is a canonical system of coordinates on $T^*\mathbf{T}^2$. The degenerations of the Liouville family include the rotationally symmetric ($f \equiv const.$) and flat ($f, g \equiv const.$).

The Liouville family is obtained from two uncoupled mechanical oscillators with periodic potentials,

$$G = \frac{1}{2}(p_x^2 + p_y^2) + a(x) + b(y), \tag{34}$$

on an energy level $E = \alpha + \beta > \max a + \max b$ such that $f = \alpha - a, g = \beta - b$. The Maupertuis principle states that orbits of the Hamiltonian flow of G on the energy level $\{G = E\}$ are orbits of the Hamiltonian flow of H up to a change in time along the orbit. The complete integrability of G is explained in Sections 2.1c and 2.3.

It is a remarkable fact that the Liouville family exhausts the list of known completely integrable Riemannian Hamiltonians whose configuration space is $\mathbf{T}^2$. Indeed, in 1989, Fomenko conjectured that these are the *only* examples possible when the second integral in polynomial-in-momenta [11]. Most recently, in 2012, Kozlov, Denisova and Treschëv reiterate Fomenko's conjecture ([12], p. 908).

Let us note that it is a well-known fact that, if the first integral F is real- analytic, then $F = \sum_{N\geq 0} F_N$ where each term F_N is polynomial-in-momenta with real-analytic coefficients, homogeneous and of degree N and since $\{H, F_N\}$ is polynomial-in-momenta, homogeneous and of degree $N+1$, each graded piece of F is a first integral. So, there is no loss in generality in restricting attention to polynomial-in-momenta first integrals–and, indeed, a slight increase in generality because the coefficients of the polynomial-in-momenta first integral are not assumed to be real-analytic.

In [13, 14], Kozlov and Denisova prove that if, when (x, y) are isothermal coordinates, and

$$H = \frac{1}{2\Lambda}(p_x^2 + p_y^2), \tag{35}$$

with the conformal factor Λ a trigonometric polynomial, then the existence of a second independent first integral that is polynomial-in-momenta implies that H is Liouville.

In [12], Denisova, Kozlov and Treschëv prove that, if one only assumes Λ is smooth, then H has no irreducible polynomial-in-momenta first integral F that is of degree 3 or 4 that is independent of H. Mironov separately proves the non-existence of F of degree 5, but as noted in ([12], p. 909), Λ satisfies an extra unstated hypothesis [15]. The line of attack used in these papers is pioneered in [16], where Kozlov and Treschëv introduce the notion of the *spectrum* $S \subset 2n\mathbf{Z}^2$ *of the function* Λ as the support of the Fourier transform of Λ. This spectrum is finite iff Λ is a trigonometric polynomial; Denisova and Kozlov prove that, in this case, any first integral of H is dependent on H unless the spectrum S is contained in a pair of orthogonal lines through (0, 0), in which case H is Liouville and has a second independent first integral that is quadratic-in-momenta. Without the hypothesis that S is finite, the problem becomes significantly more delicate. The bulk of [12], for example, is devoted to a study of solutions to a PDE that characterizes the first integral F by means of Fourier analysis.

An alternative approach, due to Bialy and Mironov, is to observe that the equation $\{H, F\} = 0$ coupled with the hypothesis that F is polynomial-in-momenta of degree N implies that when we write F as

$$F = \sum_{j=0}^{N} a_j(x, y) p_x^{N-j} p_y^j \tag{36}$$

then the coefficients $a_0, \ldots, a_n$ satisfy a *semi-linear* PDE [17, 18]. Indeed, there is a system of coordinates (τ, v) on $\mathbf{T}^2$ such that, when F is written in the adapted canonical coordinates as $F = \sum_{j=0}^{N} u_j(\tau, v)(p_v/g)^j p_r^{(N-j)}$ then this equation is of the form

$$u_v + T(u)u_r = 0 \tag{37}$$

where $u_0 = 1, u_1 = g, u = (u_1, u_2, \ldots, u_N)$ and

$$T(u)ij = \begin{cases} u_{i+1} & \text{if } j = i+1, \\ (i+1)u_{i+1} - (N-1-i)u_{i-1} & \text{if } j = 1, \\ 0 & \text{otherwise,} \end{cases} \tag{38}$$

where we adopt the convention that $u_{\_1} = u_{N+1} \equiv 0$.

A standard technique to solve a quasi-linear PDE like (37) is to diagonalize it, that is, to find Riemann invariants, so that it is equivalent to

$$r_v + \Delta(r) r_r = 0 \quad \text{where} \quad \Delta(r) = \mathrm{diag}(\delta_1(r), \ldots, \delta_N(r)), \quad r = (r_1, \ldots, r_N). \tag{39}$$

To find Riemann invariants, Bialy and Mironov employ the following trick: let $p_v = g\cos(\theta)$, $p_r = \sin(\theta)$ parameterize cotangent fibres of $H^{-1}(\frac{1}{2})$. The invariance condition $\{H, F\} = 0$ translates to $F_v g^{-1}\cos(\theta) + F_\tau \sin(\theta) = 0$ along the locus where $F_\theta = 0$, i.e. where dF and dH are co-linear. If one supposes that $\theta_i = \theta_i(\tau, v), i = 1, \ldots, N$, is a smooth parameterization of the critical-point set, then the critical values $r_i = F(r, v, \theta_i(r, v))$ are Riemann invariants with $\delta_i = g(\tau, v) \times \tan(\theta_i)$. Of course, the main problem is to determine the relationship between the Liouville foliation–the singular foliation of $T^*\mathbf{T}^2$ by the Liouville tori and their degenerations–and the system 39.

In ([18], Theorems 1 and 2), Bialy and Mironov prove that if $N \leq 4$, then in any region where a multiplier δ_i is non-real, the metric is Liouville. One can view the result of Bialy and Mironov as a partial confirmation of Fomenko's conjecture and an important step toward resolving that conjecture.

The key step in Bialy and Mironov's proof is to show that, in any region where δ_i is non-real, the imaginary part of the Riemann invariant r_i satisfies an elliptic PDE. It appears that the properties of this PDE are key to proving stronger results.

Problem 3.4. *Extend Bialy and Mironov's work to show that there are no regions where any multiplier δ_i is non-real on $\mathbf{T}^2$, i.e. show that (39) is a hyperbolic system.*

There is good reason to believe that the multipliers δ_i are always real. When $\emptyset \not\subset \delta_i^{-1}(\mathbf{C} \backslash \mathbf{R}) \not\subset \mathbf{T}^2$, Bialy and Mironov prove that the Riemann invariant r_i is real and constant, say $r_i = s_i$. This implies that the common level set $F^{-1}(s_i) \cap H^{-1}(\frac{1}{2})$, a subset of the complexified cotangent bundle $T^*_{\mathbf{C}}\mathbf{T}^2$, has a tangent with the fibres of $T^*_{\mathbf{C}}\mathbf{T}^2$ on an open set. That picture is dramatically at odds with the real picture, where the tangency can occur along a one-cycle at most. Because of this, it seems likely that there is a geometric proof of Problem 3.4.

Hyperbolicity of Eq. (39) has additional meaning. As the previous paragraph alluded to, the points where $F_\theta = 0$ are the critical points of the canonical projection map $\pi : T^*\mathbf{T}^2 \to \mathbf{T}^2$ restricted to a common level $F^{-1}(r) \cap H^{-1}\left(\frac{1}{2}\right)$. Such tori necessarily bound a solid torus in $H^{-1}\left(\frac{1}{2}\right)$ and are not minimizing. Based on Fomenko's conjecture, it is expected that these solid tori must be quite rigid in a well-defined sense: in homology, they should generate at most two transverse subgroups of $H_1(T^*\mathbf{T}^2)$.

There is an alternative approach to Fomenko's conjecture that is based on topological entropy. In a series of papers based on Glasmachers dissertation results, Glasmachers and Knieper study Riemannian Hamiltonians on $T^*\mathbf{T}^2$ with zero topological entropy [19, 20]. They prove the closure of one of the above-mentioned solid tori is a union of one or two closed, minimizing geodesic orbits and their stable and unstable manifolds ([20], Theorem 3.7c).[3] The picture that emerges from their work is that there is a family of minimizing closed geodesics of the same homology class, and their stable and unstable manifolds, which bound a family of invariant solid tori. Bialy [5] describes the boundary of this set as a separatrix chain. The projection of the separatrix chain covers $\mathbf{T}^2$. A neighbourhood of the separatrix chain in the complement is fibred by invariant Lagrangian tori that are graphs, i. e. that are a union of minimizing orbits. The multipliers δ_i, or rather the angles θ_i mentioned above, define sections of the unit cotangent bundle trapped within a separatrix chain.

Let us reformulate this as:

Problem 3.5. *Prove the vanishing of the topological entropy of the geodesic flow of a Riemannian Hamiltonian on $T^*\mathbf{T}^2$ that is completely integrable with a polynomial-in-momenta first integral F.*

In various special cases, such as when F is real-analytic or Morse-Bott, it is known that the topological entropy vanishes [21].

Finally, since topological entropy is an important invariant in the study of these systems, let us state a number of problems that are directly relevant to the preceding discussion. If one assumes Fomenko's conjecture is true and that the Liouville family of Riemannian Hamiltonians equals the set of completely integrable Riemannian Hamiltonians on $\mathbf{T}^2$, then it should be true that

Problem 3.6. *The topological entropy of a non-Liouville Riemannian Hamiltonian on $T^*\mathbf{T}^2$ is positive.*

Glasmachers and Knieper [20, 19] have studied the structure of geodesic flows with zero topological entropy on $T^*\mathbf{T}^2$. The picture that emerges is the phase portrait looks remarkably like that of an integrable system. It seems likely that their results admit a strengthening: in particular, they are unable to determine the number of primitive homology classes represented by non-minimizing geodesics (for Liouville metrics, this is at most 4).

On the other hand, it is known, from results of Contreras, Contreras and Paternain and Knieper and Weiss that an open and dense set of Riemannian Hamiltonians have positive

[3] Although the minimizing orbits have stable and unstable manifolds, it is not suggested that they are hyperbolic.

topological entropy [22–24]. In the case of this particular problem, the natural point of departure is to look at Riemannian Hamiltonians that are close to Liouville, i.e. where the conformal factor in (35) is of the form

$$\Lambda_\epsilon = \Lambda_0 + \epsilon\Lambda_1 + O(\epsilon^2) \quad (40)$$

where Λ_0 is Liouville-and has no $\mathbf{T}^1$ symmetry–and Λ_ϵ is not Liouville for all $\epsilon \neq 0$. Based on the study in [25, 26] of the phase portrait of such systems, it should be possible to prove that the perturbed flow develops transverse homoclinic points.

3.3. The 2-sphere

The unit two-dimensional sphere $\mathbf{S}^2 \subset \mathbf{R}^3$ admits a completely integrable geodesic flow. Indeed, the geodesic flow of an ellipsoid is also completely integrable with the second integral of motion that is, in general, a quadratic form in the momenta.

The fundamental problem is to describe the moduli space of completely integrable Hamiltonians on $T^*\mathbf{S}^2$. The sub-problem of describing the integrable Riemannian (resp. natural or mechanical) Hamiltonians H has received wide-spread attention. When H is Riemannian, the most common approach is to assume the second integral F is polynomial-in-momenta, and without loss of generality, homogeneous. If the degree of F is fixed, then the problem of determining H&F is reducible to a non-linear PDE in the coefficients of F. When the degree is 1, the first integral F is a momentum map of a $\mathbf{T}^1$ isometry group (see below). When the degree is 2, then the Hamiltonian is *Liouville*, a classical result due to Darboux *c.f.* [27]. In degree 3, there is the well-known case due to Goryachev-Chaplygin, and more recent cases due to Selivanova, Dullin and Matveev and Dullin, Matveev and Topalov and Valent [28–33]. In degree 4, Selivanova and Hadeler & Selivanova have produced a family of examples using the results of Kolokol'tsov [34, 27]. Beyond degree 4, Kiyohara has provided a construction of a smooth Riemannian metric H with an independent first integral F of degree k for any $k \geq 1$. In this construction, the metric H depends on a functional modulus, and so for each k, the set is infinite dimensional [35].

3.4. Super-integrable systems with a linear-in-momenta first integral

Let us review the work of Matveev and Shevchishin in more detail [36]. These authors impose an additional formal constraint that the metric possess one first integral that is linear-in-momenta. In conformal coordinates (x, y) where $H = \frac{1}{2}c(x)(p_x^2 + p_y^2)$, the existence of a cubic integral is reduced to a second-order ODE involving c.

From a geometric perspective, it is more natural to introduce coordinates adapted to the isometry group. That is, the existence of a linear-in-momenta first integral is equivalent to the existence of an isometry group containing $\mathbf{T}^1$. The action of $\mathbf{T}^1$ on $\mathbf{S}^1$ induces a cohomogeneity-1 structure. The fixed set of the $\mathbf{T}^1$ action is a set of points $\{p_-, p_+\}$ which are equidistant along any minimal geodesic; and the principal $\mathbf{T}^1$-orbits are orthogonal to these geodesics. If $\gamma : [-T, T] \to \mathbf{S}^2$ is a minimal geodesic such that $\gamma(\pm T) = p_\pm$, then we can let (r, θ) be 'polar' coordinates adapted to this structure. The Hamiltonian H and polynomial-in-momenta integral F can be written in the adapted coordinates as

$$H = \frac{1}{2}\left(p_r^2 + s(r)p_\theta^2\right), F = e^{iv\theta} \times \sum_{j=0} a_j(r) p_r^j p_\theta^{N-j}, \tag{41}$$

where $v \in \mathbf{Z}$, $3 \leq N$ is a positive integer and the coefficients a_j are to be determined. The equation $\{H, F\} \equiv 0$ is equivalent to a differential system that couples the coefficients $a_0, \ldots, a_N$, s and an anti-derivative S of vs:

$$dS = vs\,\mathrm{d}r, \tag{42a}$$

$$\mathrm{d}a_j = \frac{1}{2}(N+2-j)a_{j-2}\,\mathrm{d}s - a_{j-1}\,\mathrm{d}S; (j = 0, \ldots, N), \tag{42b}$$

$$\mathrm{d}s = 2va_N/a_{N-1}\,\mathrm{d}r \tag{42c}$$

where $a_{-2} = a_{-1} = 0$. It is clear that the general solution of (42b), without the compatibility condition (42c), is obtained via repeated quadratures of products of s and S. The compatibility condition distinguishes those solutions which may arise from (41). The behaviour of s at $r = \pm T$ ultimately determines whether the solution obtained arises from a $\mathbf{T}^1$-invariant Riemannian Hamiltonian H and an independent first integral F on T^*S^2.

In case $N = 3$, the differential system reduces to a third-order nonlinear ODE similar to that studied by Chazy, in his generalization of the Painlêvé classification ([37], Eq. (6)). Based on the work of Matveev and Shevchishin [36], we know the solutions to this equation are real-analytic and define a parameterized family of super-integrable Riemannian metrics with cubic-in-momenta first integral. The latter authors do not solve the ODE explicitly.

***Problem* 3.7.** *Solve the* $N = 3$ *case of the differential system (42).*

It appears to the author that this differential system may be soluble via hypergeometric functions. A successful resolution to the $N = 3$ case will naturally lead to the higher degree cases, which appear to be somewhat more involved.

***Problem* 3.8.** *Solve the higher degree cases of the differential system (42).*

3.5. Super-integrable systems with a higher degree first integral

The author believes that the differential system 42 provides the key to understanding the subspace of super-integrable Riemannian Hamiltonians which admit a cohomogeneity-1 structure. Super-integrability alone does not imply the existence of such a cohomogeneity-1 structure. Without this additional hypothesis, there is very little known. Indeed, the extremely valuable construction of Kiyohara is the only construction that provides a smooth Riemannian Hamiltonian with a polynomial-in-momenta first integral of degree $N > 3$–super-integrable or not [35, 38].

Let us explain Kiyohara's construction in some detail. Let H_0 be the Riemannian Hamiltonian of the standard unit sphere in $\mathbf{R}^3$. Let F_0, F_1 be linear-in-momenta first integrals of H_0 that are linearly independent and let $l \geq k \geq 1$ be integers such that $N = k + l \geq 3$. Define a polynomial-in-momenta first integral $G_0 = F_0^k F_1^l$. For almost all $q \in \mathbf{S}^2$, the functions G_0, $H_0|T_q^*\mathbf{S}^2$ are dependent along two distinct lines through 0; this defines a pair of mutually transverse line bundles $L_\pm^*$ over $\mathbf{S}^2 \backslash \{p_0^\pm, p_1^\pm\}$. The excluded, singular set consists of the anti-podal points $p_j^\pm$ where F_j vanishes identically on the fibre (equivalently, the corresponding Killing field vanishes). This pair of line bundles provides a branched double covering

$$\Phi : \mathbf{T}^2 = \mathbf{R}^2/2\pi\mathbf{Z}^2 \rightarrow \mathbf{S}^2 \tag{43}$$

with simple branch points at $\{p_0^\pm, p_1^\pm\} = \Phi(\pi\mathbf{Z}^2)$. The line bundles $L_\pm^*$ pullback to the line bundles R dx_j on $\mathbf{T}^2 = \{(x_1, x_2) \bmod 2\pi\mathbf{Z}\}$. Kiyohara shows that in these coordinates, the pull-back of the function r which measures the time along the unique geodesic γ through $\{p_0^\pm, p_1^\pm\}$ (see **Figure 4**) satisfies the second-order PDE

$$\frac{\partial^2 r}{\partial x_1 \partial x_2} + \frac{1}{B_1 + B_2} \frac{\partial B_1}{\partial x_2} \frac{\partial r}{\partial x_1} + \frac{1}{B_1 + B_2} \frac{\partial B_2}{\partial x_1} \frac{\partial r}{\partial x_2} = 0 \tag{44}$$

where B_1 and B_2 are functions that describe the line bundles $L_\pm$ in terms of the basis $\{\mathrm{d}r, \sin(r)\, \mathrm{d}\theta\}$.

Kiyohara writes a function $R = r_0 + r$ where r_0 is the solution to (44) given by $\Phi^* r$ and r is a solution of (44) with C^2 small boundary conditions satisfying

$$r(s, 0) = u_1(s), r(0, s) = u_2(s), \tag{45a}$$

$$\text{where} \quad u_i(s) = u_i(-s) = u_i(\pi - s), \quad \text{forall } i, s, \text{ and} \tag{45b}$$

$$u_i([-\epsilon, \epsilon]) = 0. \tag{45c}$$

Then, by means of this perturbed function R, Kiyohara writes down an explicit formula for the perturbed Riemannian Hamiltonian H and polynomial-in-momenta first integral F. The condition for the Poisson bracket $\{H, F\}$ to vanish is shown to reduce to the satisfaction of Eq. (44) by R for the *given* values of B_1 and B_2 (this legerdemain is the real trick that makes the construction work).

Condition (45b) ensures that R factors through Φ to a function on $\mathbf{S}^2$, while the condition (45c) ensures that R is C^∞ on $\mathbf{S}^2$ and coincides with r on a neighbourhood of the branch set $\{p_0^\pm, p_1^\pm\}$ (hence that H and F coincide with H_0 and F_0, respectively, on a neighbourhood of the cotangent fibres of the branch set).

Let us now state several problems related to Kiyohara's construction. First, Kiyohara's vanishing condition on the boundary values (45c) is used to deduce the Riemannian Hamiltonians are not real-analytic. Since all the remaining constructions involve real-analytic data, this serves to show his examples are genuinely different.

Problem 3.9. *Does Kiyohara's construction extend to real-analytic boundary conditions u_1, u_2 that satisfy (45b) ? Do these real-analytic metrics include other known cases?*

http://dx.doi.org/10.5772/66147

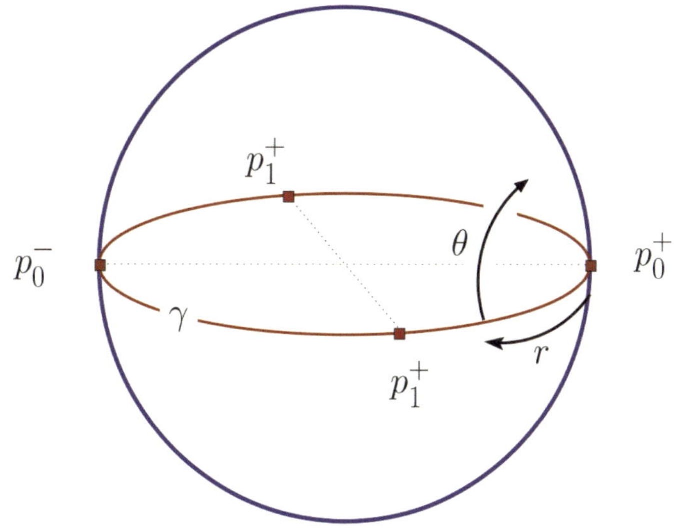

Figure 4. Kiyohara's construction. The zero set of the pair of Killing fields determines the equatorial geodesic γ. A choice of zeros $\{p_0^+, p_1^+\}$ determines the polar coordinate system (r, θ).

In particular, the obtained metrics are unlikely to have a $\mathbf{T}^1$ isometry group, so the question is really whether the known examples in degree 3 and 4 are obtainable via this construction [12–14, 28–30, 34, 39, 40].

Second, Kiyohara's construction produces a polynomial-in-momenta first integral F factors as $A_0^l A_1^m$ where A_i are linear-in-momenta functions. It is clear that the reducibility of the first integral F is forced by the desire to use a very simple branched covering.

Problem 3.10. *Is reducibility of the first integral F necessary?*

It ought to be fruitful to ask three related questions. The reducibility of F is very special, with just two distinct factors.

Problem 3.11. *Is it possible to extend Kiyohara's construction so that the polynomial-in-momenta first integral* F *has more than 2 distinct linear factors?*

It would be natural to try to extend the construction to the case where the zeros all lie on the same geodesic γ. More generally, one might attempt to mirror Kiyohara's construction but in a more abstract way: start with a simple ramified covering $\Phi : \sum \to \mathbf{S}^2$ with a branch set $Y \subset \mathbf{S}^2$. Let F_0 be a product of linear first integrals of H_0 that vanishes identically on $T_Y^*\mathbf{S}^2$ and not elsewhere. The stumbling block is that we need to clarify the intrinsic geometric meaning of the PDE that governs the perturbed systems (44).

Problem 3.12. *Describe in explicit terms the third, independent first integral of H that is of least degree.*

Kiyohara proves in his paper that H is super-integrable (he proves the geodesic flow is 2π-periodic, in fact), but that proof does not proceed by finding this third first integral.

3.6. Three-dimensional configuration spaces

In comparison to the wealth of results and examples for surfaces that were surveyed above, comparatively little is known about the three-dimensional analogues. Tăĭmanov tells us that if the Tonelli Hamiltonian is completely integrable with real-analytic first integrals, then the three-dimensional configuration space Σ has a finite covering $\hat{p} : \hat{\Sigma} \to \Sigma$ such that the fundamental group $\pi_1(\hat{\Sigma})$ is abelian and of rank at most 3 [41–43]. Based on the resolution of the Poincaré conjecture, this result implies that, up to finite covering the only such configuration spaces are

$$\mathbf{S}^3, \qquad \mathbf{S}^2 \times \mathbf{T}^1 \qquad \text{or} \qquad \mathbf{T}^3. \tag{46}$$

The author generalized Kozlov's result on surfaces to three-manifolds. In this result, if the Tonelli Hamiltonian is completely integrable and the singular set is topologically tame, then Tăĭamanov's list extends to include those three-manifolds Σ such that $\pi_1(\Sigma)$ is almost solvable (equivalently, due to the resolution of the geometrization conjecture, Σ admits either a *Nil* or *Sol* geometry) [44]. Both results are sharp, like Kozlov's, in the sense that all such admissible configuration spaces admit a geometric structure and the Riemannian Hamiltonian of such a structure is completely integrable with first integrals of the requisite type [45, 46].

There are a large number of questions that this strand of research has opened. Let us sketch a few.

3.7. The 3-sphere

The case of $\mathbf{S}^3$ is perhaps best understood. It has been known since Jacobi proved the complete integrability of the geodesic flow of an ellipsoid via separation of variables, that the Liouville family of metrics on $\mathbf{S}^3$ is completely integrable. These systems possess three independent quadratic-in-momenta first integrals.

Based on the analogous problem for the two-sphere,

Problem 3.13. *Describe the structure of the super-integrable Riemannian Hamiltonians on* $\mathbf{S}^3$.

Researchers who specialize in super-integrable classical and quantum systems have developed tools for constructing and classifying super-integrable systems *c.f.* [47–49]. Unfortunately, some key ingredients in these constructions lead to systems with singularities.

The first method is based on the cohomogeneity-1 structure of $\mathbf{S}^3$ with the group $G = \mathrm{SO}(3)$ acting as the linear isometry group of $\mathbf{R}^3 \subset \mathbf{R}^4$. If one represents

$$\mathbf{S}^3 = \{(x, r) | x \in \mathbf{R}^3, r \in \mathbf{R}, |x|^2 + |r|^2 = 1\}, \tag{47}$$

then we see that G acts freely on $T^*\mathbf{S}^3 \backslash T^*_F\mathbf{S}^3$ where $F = \{(0, \pm 1)\}$ is the fixed-point set of the G-action on $\mathbf{S}^3$. This is enough to see that any G-invariant Hamiltonian on $T^*\mathbf{S}^3$ is non-commutatively integrable (analogous to the same fact for $\mathbf{S}^2$). If $K : \mathrm{so}(3)^* \to \mathrm{R}$ is a positive-definite quadratic form, and $\Psi : T^*\mathbf{S}^3 \to \mathrm{so}(3)^*$ is the momentum map of the SO(3)-action, then an invariant Riemannian Hamiltonian can be written as

$$H = \frac{1}{2}p_r^2 + \frac{1}{2}s(r)\Psi^*K, \tag{48}$$

for some function $s > 0$ such that $s \times (1 \pm r)^2 \to const. \neq 0$ as $r \to \mp 1$.

If one employs the *ansatz* of Matveev & Shevchishin (*c.f.* Section 3.3), one would like to find first integrals that are polynomial-in-momenta of the form

$$F = \sum_{j=0}^{N} b_j(x,r)\, p_r^j\, \Psi^* \eta_{N-j} \tag{49}$$

where $\eta_{N-j} : \mathrm{so}(3)^* \to \mathbf{R}$ is a homogeneous polynomial of degree $N-j$. In (41), the pre-factor $\exp(iv\theta)$ appears to ensure that the coefficients of the first integral F are common eigenfunctions of the Casimir $\Delta_{\mathbf{S}^1} = \frac{\partial^2}{\partial\theta^2}$ parameterized by r. In the current case, the *ansatz* suggests that the coefficients b_j should factor as $\phi_\lambda(\theta)a_j(r)$ where ϕ_λ is an eigenfunction of the Casimir $\Delta_{\mathbf{S}^2}$ with eigenvalue λ and $\theta = x/|x|$.

Problem 3.14. *Extend the construction sketched, above to higher dimensional spheres.*

Author details

Leo T. Butler

Address all correspondence to: leo.butler@ndsu.edu

Department of Mathematics, North Dakota State University, Fargo, ND, USA

References

[1] H. Hancock. *Elliptic integrals*. Dover Publications, Inc., New York, 1958.

[2] F. Magri and C. Morosi. A Geometrical Characterization of Integrable Hamiltonian Systems Through the Theory of Poisson-Nijenhuis Manifolds. Department of Mathematics, University of Milan, 1984.

[3] A. S. Miščcenko and A. T. Fomenko. A generalized Liouville method for the integration of Hamiltonian systems. *Funkcional. Anal. Priložen.*, 12(2):46–56, 96, 1978.

[4] B. A. Kupershmidt. *Elements of Superintegrable Systems*, volume 34 of *Mathematics and its Applications*. D. Reidel Publishing Co., Dordrecht, 1987. Basic techniques and results.

[5] M. Bialy. Integrable geodesic flows on surfaces. *Geom. Funct. Anal.*, 20(2):357–367, 2010.

[6] V. V. Kozlov. Topological obstacles to the integrability of natural mechanical systems. *Dokl. Akad. Nauk SSSR*, 249(6):1299–1302, 1979.

[7] L. T. Butler. A note on integrable mechanical systems on surfaces. *DCDS-A*, 34(8):1873–1878, 2014.

[8] Y. Long. Collection of problems proposed at International Conference on Variational Methods. *Front. Math. China*, 3(2):259–273, 2008.

[9] A. Katok. Lyapunov exponents, entropy and periodic orbits for diffeomorphisms. *Inst. Hautes Études Sci. Publ. Math.*, (51):137–173, 1980.

[10] A. Manning. Topological entropy for geodesic flows. *Ann. Math. (2)*, 110(3):567–573, 1979.

[11] A. T. Fomenko. Symplectic topology of completely integrable Hamiltonian systems. *Uspekhi Mat. Nauk*, 44(1(265)):145–173, 248, 1989.

[12] N. V. Denisova, V. V. Kozlov, and D. V. Treshchëv. Remarks on polynomial integrals of higher degree for reversible systems with a toral configuration space. *Izv. Ross. Akad. Nauk Ser. Mat.*, 76(5):57–72, 2012.

[13] V. V. Kozlov and N. V. Denisova. Symmetries and topology of dynamical systems with two degrees of freedom. *Mat. Sb.*, 184(9):125–148, 1993.

[14] V. V. Kozlov and N. V. Denisova. Polynomial integrals of geodesic flows on a two-dimensional torus. *Mat. Sb.*, 185(12):49–64, 1994.

[15] A. E. Mironov. Polynomial integrals of a mechanical system on a two-dimensional torus. *Izv. Ross. Akad. Nauk Ser. Mat.*, 74(4):145–156, 2010.

[16] V. V. Kozlov and D. V. Treshchëv. The integrability of Hamiltonian systems with configuration space a torus. *Mat. Sb. (N.S.)*, 135(177)(1):119–138, 144, 1988.

[17] M. Bialy and A. E. Mironov. Rich quasi-linear system for integrable geodesic flows on 2-torus. *Discrete Contin. Dyn. Syst.*, 29(1):81–90, 2011.

[18] M. Bialy and A. E. Mironov. Cubic and quartic integrals for geodesic flow on 2-torus via a system of the hydrodynamic type. *Nonlinearity*, 24(12):3541–3554, 2011.

[19] E. Glasmachers and G. Knieper. Characterization of geodesic flows on T^2 with and without positive topological entropy. *Geom. Funct. Anal.*, 20(5):1259–1277, 2010.

[20] E. Glasmachers and G. Knieper. Minimal geodesic foliation on T^2 in case of vanishing topological entropy. *J. Topol. Anal.*, 3(4):511–520, 2011.

[21] G. Paternain. Entropy and completely integrable Hamiltonian systems. *Proc. Am. Math. Soc.*, 113(3):871–873, 1991.

[22] G. Contreras. Geodesic flows with positive topological entropy, twist maps and hyperbolicity. *Ann. Math. (2)*, 172(2):761–808, 2010.

[23] G. Contreras-Barandiaran and G. P. Paternain. Genericity of geodesic flows with positive topological entropy on S^2. *J. Differ. Geom.*, 61(1):1–49, 2002.

[24] G. Knieper and H. Weiss. C^∞ genericity of positive topological entropy for geodesic flows on S^2. *J. Differ. Geom.*, 62(1):127–141, 2002.

[25] E. N. Selivanova. Classification of geodesic flows of Liouville metrics on a two-dimensional torus up to topological equivalence. *Mat. Sb.*, 183(4):69–86, 1992.

[26] E. N. Selivanova and A. M. Stepin. On the dynamic properties of geodesic flows of Liouville metrics on a two-dimensional torus. *Tr. Mat. Inst. Steklova*, 216(Din. Sist. i Smezhnye Vopr.):158–175, 1997.

[27] V. N. Kolokol'tsov. Geodesic flows on two-dimensional manifolds with an additional first integral that is polynomial with respect to velocities. *Izv. Akad. Nauk SSSR Ser. Mat.*, 46 (5):994–1010, 1135, 1982.

[28] E. N. Selivanova. New examples of integrable conservative systems on S^2 and the case of Goryachev-Chaplygin. *Comm. Math. Phys.*, 207(3):641–663, 1999.

[29] K. P. Hadeler and E. N. Selivanova. On the case of Kovalevskaya and new examples of integrable conservative systems on S^2. *Regul. Chaot. Dyn.*, 4(3):45–52, 1999.

[30] H. R. Dullin, V. S. Matveev, and P. I. Topalov. On integrals of the third degree in momenta. *Regul. Chaot. Dyn.*, 4(3):35–44, 1999.

[31] H. R. Dullin and V. S. Matveev. A new integrable system on the sphere. *Math. Res. Lett.*, 11 (5–6):715–722, 2004.

[32] H. R. Dullin and V. S. Matveev. A new natural Hamiltonian system on T^*S^2 admitting an integral of degree 3 in momenta. In *Global Analysis and Applied Mathematics*, volume 729 of *AIP Conf. Proc.*, pp. 141–146. Amer. Inst. Phys., Melville, NY, 2004.

[33] G. Valent. On a class of integrable systems with a cubic first integral. *Comm. Math. Phys.*, 299(3):631–649, 2010.

[34] E. N. Selivanova. New families of conservative systems on S^2 possessing an integral of fourth degree in momenta. *Ann. Global Anal. Geom.*, 17(3):201–219, 1999.

[35] K. Kiyohara. Two-dimensional geodesic flows having first integrals of higher degree. *Math. Ann.*, 320(3):487–505, 2001.

[36] V. S. Matveev and V. V. Shevchishin. Two-dimensional superintegrable metrics with one linear and one cubic integral. *J. Geom. Phys.*, 61(8):1353–1377, 2011.

[37] J. Chazy. Sur les equations diflarentielles du troisieme ordre et d'ordre superieur dont l'integrale generale a ses points critiques fixes. *Acta Math.*, 34(1):317–385, 1911.

[38] K. Kiyohara. Periodic geodesic flows and integrable geodesic flows. *Sugaku*, 56(1):88- 98, 2004.

[39] S. Gravel. Hamiltonians separable in Cartesian coordinates and third-order integrals of motion. *J. Math. Phys.*, 45(3):1003–1019, 2004.

[40] N. V. Denisova and V. V. Kozlov. Polynomial integrals of reversible mechanical systems with a two-dimensional torus as configuration space. *Mat. Sb.*, 191(2):43–63, 2000.

[41] I. A. Taĭmanov. Topological obstructions to the integrability of geodesic flows on nonsimply connected manifolds. *Izv. Akad. Nauk SSSR Ser. Mat.*, 51(2):429–435, 448, 1987.

[42] I. A. Taimanov. Topological properties of integrable geodesic flows. *Mat. Zametki*, 44 (2):283–284, 1988.

[43] I. A. Taĭmanov. Topology of Riemannian manifolds with integrable geodesic flows. *Trudy Mat. Inst. Steklov.*, 205(Novye Rezult. v Teor. Topol. Klassif. Integr. Sistem):150–163, 1994.

[44] L. T. Butler. Invariant fibrations of geodesic flows. *Topology*, 44(4):769–789, 2005.

[45] A. V. Bolsinov and I. A. Taimanov. Integrable geodesic flows with positive topological entropy. *Invent. Math.*, 140(3):639–650, 2000.

[46] L. T.Butler. New examples of integrable geodesic flows. *Asian J. Math.*, 4(3):515–526, 2000.

[47] E. G. Kalnins, W. Miller, Jr., and S. Post. Two-variable Wilson polynomials and the generic superintegrable system on the 3-sphere. *SIGMA Symmetry Integrability Geom. Methods Appl.*, 7:Paper 051, 26, 2011.

[48] E. G. Kalnins, J. M. Kress, and W. Miller, Jr. Fine structure for 3D second-order superintegrable systems: three-parameter potentials. *J. Phys. A*, 40(22):5875–5892, 2007.

[49] E. G. Kalnins, J. M. Kress, and W. Miller, Jr. Extended Kepler-Coulomb quantum superintegrable systems in three dimensions. *J. Phys. A*, 46(8):085206, 28, 2013.

Section 2

Applications

Chapter 4

Closure Models for Lagrangian Gas Dynamics and Elastoplasticity Equations in Multimaterial Cells

Yury Yanilkin

Additional information is available at the end of the chapter

http://dx.doi.org/10.5772/66858

Abstract

Mixed cells (multicomponent cells) emerging in the development of Lagrangian-Eulerian (ALE) or Eulerian numerical techniques for solving the gas dynamics and elastoplasticity equations in multicomponent media contain either interfaces between materials or a mixture of materials. There is a problem of correctly approximation of the equations in such cells and the ALE code accuracy and performance depend on how the problem is resolved. Many approximation methods use the equation splitting into two stages, one of which consists in solving a given equation in Lagrangian variables. If mixed cells are simulated, the system of equations describing the gas dynamics and elastoplasticity is unclosed and there is a need to introduce additional closure relations that will allow determining the thermodynamic parameters of components using the available data for the mixture of components, as a whole. The chapter presents a review of the equation closure methods and results of the methods verification using several test problems having exact solutions.

Keywords: ALE method, mixed cell, closure model, numerical simulation, verification

1. Introduction

Mixed cells in arbitrary Lagrangian-Eulerian (ALE) or Eulerian methods contain interfaces between different materials or a mixture of materials. In the next section, we will not distinguish these two methods, considering that both methods solve the advection equation, including the vicinity of mixed cells. Most of these methods use a two-stage approximation of equations. The first stage considers gas dynamics or elastoplasticity equations without convective terms. The convective transfer comes into play at the second stage. Among many similar methods, we consider only the ALE methods that contain Lagrangian gas dynamics and elastoplasticity in the pure form, and the problem of mixed cells at that particular stage is the subject of research reported here. Note that mixed cells can be present even in purely

Lagrangian techniques, and the problems related to their presence should also be addressed in this case.

Here, we will generally use the term "Lagrangian gas dynamics" (or simply "gas dynamics"), bearing in mind that, in the case of elastoplasticity, this will also involve equation terms related to the stress tensor deviator. Historically, several approaches to the problem of mixed cells in gas dynamics associated with materials distinction in such cells have been considered. In this chapter, we consider only the single-velocity model of matter. The major approach that has become predominant these days uses complete thermodynamic distinction of materials.[1] Next, we will use the term "material" meaning that, mathematically, an interface can also divide identical materials; moreover, one of the materials can be vacuum and/or a perfectly rigid body.

Thermodynamic parameters in gas dynamics include density, internal energy, and pressure. If other processes are modeled, the number of parameters increases; for example, for elastoplasticity, additional parameters will include components of the stress tensor deviator. In addition to thermodynamic parameters, volume fractions of constituent materials are introduced in each mixed cell that can be used to determine the geometric location of the interface inside a mixed cell, which is used in some models.[2]

This approach to materials identification allows one to model mixed cells containing not only contacting but also intermingled materials. When mixed cells are used for gas dynamics equations, additional closing relations are needed, which in fact define the interaction of materials inside the cell (the subcell interaction). Most of the known models manage with information about volume (or mass) fractions of the materials and their thermodynamic states [12–26]. Such models can be divided into two classes according to the number of computational stages involved.[3]

The first class of models is based on introducing closure models at a single stage, while the second class includes two-stage models, in which the second stage is in fact complementary to the first one and involves additional interaction between materials inside a mixed cell (so-called subcell interaction).

Next, we often use the terms "model" and "method" without distinction. One should note here that a method is understood to be an algorithm implemented in the form of a program and based on some physical model.

---

[1]Early in the development of Eulerian methods, a smaller number of parameters have been used to identify the materials; for example, in [1, 2], mass fractions of the materials and average energy of matter were employed. Accordingly, other closing relations were used, the required number of which in this case is plus one compared with the complete materials distinction. The models thus considered include the "isobar-isothermal" and the "isobar-isodQ" models, which, although successful in some respect, in the general case failed to deliver acceptable accuracy of results.

[2]The problem of identifying the contact location based on the material volume fractions is beyond the scope of this study; it is a separate problem discussed in dedicated studies (see, e.g. [1, 3–11]).

[3]Our classification and description of models is limited to the case of two materials in a cell, although many formulas mentioned in this chapter are also suitable for their larger number. For this reason, some models developed specifically for the case of several materials in a cell are left beyond the scope of our review.

http://dx.doi.org/10.5772/66858

Basic *single-stage* closure methods include the following:

1. Method based on the model of equal pressures of constituent materials (the P method) [12],
2. Tipton's method [13],
3. Delov's method based on the acoustic Riemann solver [14] (this approach is also used in the DSS [22] and KSR [23] methods developed later),
4. The K&S method based on a local Riemann problem [15]. This method is not described in this work because of its impracticability, as noted in [15], but its test simulation results are given for comparison.

Note that these four models and methods developed on their basis are relaxation with respect to pressure. In a number of studies, nonrelaxation methods have been proposed, which use the following assumptions (models):

5. Equal velocity divergences ($\nabla \cdot \mathbf{u}$) of the materials [17].
6. Equal pressure changes (Δp) of the materials [18],
7. Equal velocities (Δu) of the materials behind a weak shock wave [19].

Two-stage models include the stage of subcell interactions between the materials in the nonequilibrium state; so the first stage here can only use models 5–7. This approach for closure models has been proposed independently in [20, 21]. The subcell pressure relaxation method in [20] is versatile and it is used in combination with models 5–7, denoted as the $\nabla \cdot \mathbf{u}$-PR, Δp-PR and Δu-PR (pressure relaxation) methods.

All the above-mentioned methods do not employ the contact location inside a mixed cell. However, there are methods that make essential use of the information about the contact location. A method of this kind was first proposed in [21] and then developed in the "interface-aware subscale dynamics" IA-SSD method [24, 25] for the multimaterial case. It offers a two-stage model, the first stage of which employs the $\nabla \cdot \mathrm{u}$ model. At the second stage, driven by the materials' individual pressures, the interface between the materials moves normally to it. The interface is reconstructed based on the volume fractions, and its motion is accomplished based on the solution of an acoustic Riemann problem (model 3).

Let us point out one common feature (associated with the assumptions made in the models) of all the above closure methods. Velocity in the methods is normal to the interface (definite or imaginary) irrespective of interface location relative to the vector of velocity. In fact, they are isotropic in the sense that compression (expansion) ratios of materials are assumed to be equal in all directions. One can mention a number of other closure methods that employ algorithms similar to those used in the above-mentioned models [27–33]. This property of the methods is quite acceptable for most applications, but there are problems (see below), as applied to which it results in significant errors in simulations.

In [34], anisotropic closure methods, ACM-1 and ACM-2, are proposed, which are an extension of models 5–7. They possess all the advantages of methods 5–7, which are central to the

EGAK code [35] when modeling flows, for which one can assume that they are isotropic, but have an important advantage when modeling more complex flows.

Apart from the basic closure method, mixed cells require additional relations to address the ways of pressure and artificial viscosity calculations for the whole cell and artificial viscosity calculations for the materials. Six approaches to calculate the artificial viscosity of materials are discussed in [36].

2. Finite difference approximation of elastoplasticity equations

2.1. Initial equations for multimaterial elastoplasticity

The initial set of equations solved at the Lagrangian stage for 2D elastoplastic flows is the following:

$$\frac{d\mathbf{u}}{dt} = -\frac{1}{\rho}\nabla \cdot \mathbf{T}, \tag{1}$$

$$\frac{d\rho_\xi}{dt} = -\rho_\xi \nabla \cdot \mathbf{u}_\xi, \tag{2}$$

$$\frac{d\beta_\xi}{dt} = \beta_\xi(\nabla \cdot \mathbf{u}_\xi - \nabla \cdot \mathbf{u}), \tag{3}$$

$$\frac{de_\xi}{dt} = \frac{1}{\rho_\xi} Sp(\mathbf{T}_\xi \mathbf{D}_\xi), \tag{4}$$

$$\frac{d\mathbf{r}}{dt} = \mathbf{u}. \tag{5}$$

In this set of equations: $\mathbf{u}(u_x, u_y)$ is the velocity, ρ is the density, $\mathbf{T}$ is the stress tensor, $\mathbf{D}$ is the strain rate tensor, and e is the specific internal energy, $\beta$is the volume fraction of the material ($\beta_\xi = \frac{M_\xi}{V}$), $\mathbf{r}(x,y)$ is the radius vector. The subscript ξ is the material index; also note that in the expression for the velocity divergence (or simply divergence for short) it relates to the divergence as a whole, rather than to the velocity. Bold type here and below is used to indicate the vector, tensor, and deviator.

Eq. (3) can be derived from the equation of continuity (2), in which the materials' density is substituted with its expression in the form of $\rho_\xi = \frac{M_\xi}{V_\xi}$, where M_ξ and V_ξ are the mass and volume of the materials in the Lagrangian cell. We obtain $\frac{dV_\xi}{dt} = V_\xi \nabla \cdot \mathbf{u}_\xi$ and then introduce an expression for V_ξ in terms of volume fractions $V_\xi = \beta_\xi V$ into this equation. Thus, Eq. (3) is a consequence of Eq. (2), and we give it here solely for the purpose of empathizing that the volume fractions for multimaterial matter should also be updated to t^{n+1}. In the single-material case, Eqs. (1)–(5) come down to usual Lagrangian gas dynamics equations, because Eq. (3) is not present in this case, and the quantities in other equations are written without material indices.

http://dx.doi.org/10.5772/66858

Stress and strain rate tensors are expressed as follows:

$$\mathbf{T} = \begin{bmatrix} T_{xx} & T_{xy} & 0 \\ T_{yx} & T_{yy} & 0 \\ 0 & 0 & \mathbf{T}_{\phi} \end{bmatrix}, \quad \mathbf{D} = \begin{bmatrix} d_{xx} & d_{xy} & 0 \\ d_{yx} & d_{yy} & 0 \\ 0 & 0 & d_{\phi} \end{bmatrix}. \tag{6}$$

The stress tensor is represented as a sum of the spherical part (pressure p) and the deviator $\mathbf{S}(S_{xx}, S_{yy}, S_{xy}, S_{\phi})$. Deviator components are defined by the relation $S_{ij} = T_{ij} - \delta_{ij}p$.

For the materials, we define equations of state

$$p_{\xi} = P_{\xi}(\rho_{\xi}, e_{\xi}), \tag{7}$$

and equations to express the deviator $\mathbf{S}_{\xi}$ as a function of the strain rate tensor $\mathbf{D}_{\xi}$

$$f_{\xi}(\mathbf{S}_{\xi}, \mathbf{D}_{\xi}) = 0. \tag{8}$$

The specific form of Eq. (8) is determined by the model of matter adopted.

EGAK uses decomposition in physical processes, in which the pressure-related terms are approximated at the Lagrangian gas dynamics stage, and the terms related to the stress tensor deviator, at the other stage of the computation. In the present technique, materials can be both different substances with their equations of state and vacuum.

2.2. Finite difference approximation of elastoplasticity equations

EGAK uses a quadrangular mesh with node-centered velocities and all the other quantities ($\rho_{\xi}, \beta_{\xi}, e_{\xi}, p_{\xi}, \mathbf{S}_{\xi}$) defined at cell centers and for each material individually. Also note that for the purpose of program implementation, pressure in Eqs. (1)–(5) is replaced with a sum of pressure and artificial (computational) viscosity for matter as a whole and for the materials, $p \to p + q$ and $p_{\xi} \to p_{\xi} + q_{\xi}$, respectively. Known quantities (basic variables) in Eqs. (1)–(5) in the 2D case include $u_x, u_y, \rho_{\xi}, \beta_{\xi}, e_{\xi}, p_{\xi}, \mathbf{S}_{\xi}$, and the quantities p, $\mathbf{S}$, q, q_{ξ}, $\nabla \cdot \mathbf{u}_{\xi}$, $\mathbf{D}_{\xi}$ need to be determined. In the following formulas, the subscript means discretization in space and the material number ξ in the multimaterial case, and the superscript denotes discretization in time. Cell-centered quantities are marked with a semi-integer superscript (for example, i +1/2), and the node-centered ones, with an integer superscript (i). If in a specific formula it is clear from the context that the superscripts are the same for all the quantities, they are omitted. Cell masses in Lagrangian gas dynamics remain constant in the course of calculations, so they have no temporal indexing.

Suppose that we know all the basic quantities at time t^n and that we seek to update their values at time $t^{n+1} = t^n + \tau$, where τ is the timestep chosen based on the requirement that the difference scheme should be stable (these issues are beyond the scope of this work). Let us write the difference scheme of EGAK for the multimaterial case.

First half-timestep (determination of predicted pressure)

$$\mathbf{r}^{n+1/2} = \mathbf{r}^n + \tau \cdot \mathbf{u}^n, \tag{9}$$

$$V^{n+1/2}_{i+1/2,j+1/2} = V\left(\mathbf{r}^{n+1/2}_{i+1,j+1}, \mathbf{r}^{n+1/2}_{i,j+1}, \mathbf{r}^{n+1/2}_{i+1,j}, \mathbf{r}^{n+1/2}_{i,j}\right), \tag{10}$$

$$p^n_\xi = P(\rho^n_\xi, e^n_\xi), \tag{11}$$

$$\nabla \cdot \mathbf{u}^n = \frac{(V^{n+1/2} - V^n)}{(\tau \cdot V^n)}, \tag{12}$$

$$p^{n+1/2}_\xi = p^n_\xi - \chi \cdot \rho^n_\xi \cdot (c^n_\xi)^2 \cdot \tau \cdot \overline{\nabla \cdot \mathbf{u}}^n_\xi. \tag{13}$$

In Eq. (13), $\chi = 0.6$ (this value was chosen in [17]), c^n_ξ is the speed of sound.

Full timestep

$$M_{i,j} = 0.25 \cdot \sum_\xi \left(M_{\xi,i-1/2,j-1/2} + M_{\xi,i-1/2,j-1/2} + M_{\xi,i+1/2,j-1/2} + M_{\xi,i+1/2,j+1/2}\right), \; M_\xi = \rho_\xi \cdot \beta_\xi \cdot V, \tag{14}$$

$$M_{i,j} \frac{(\mathbf{u}^{n+1}_{i,j} - \mathbf{u}^n_{i,j})}{\tau} = \left(\overline{\nabla g}^{n+1/2} + \overline{\nabla \cdot S}^n\right)_{i,j}, \tag{15}$$

$$\Rightarrow \mathbf{u}^{n+1}_{i,j} = \mathbf{u}^n_{i,j} - \left(\frac{\tau}{M_{i,j}}\right) \cdot \left(\overline{\nabla g}^{n+1/2} + \overline{\nabla \cdot S}^n\right)_{i,j}, \tag{16}$$

$$\mathbf{u}^{n+1/2} = (\mathbf{u}^n + \mathbf{u}^{n+1})/2, \tag{17}$$

$$\mathbf{r}^{n+1} = \mathbf{r}^n + \tau \cdot \mathbf{u}^{n+1}, \tag{18}$$

$$V^{n+1}_\xi = V^n_\xi + \tau \cdot V^{n+1}_\xi \overline{\nabla \cdot \mathbf{u}}^{n+1}_\xi, \tag{19}$$

$$\rho^{n+1}_\xi = \frac{M_\xi}{V^{n+1}_\xi}, \tag{20}$$

$$\beta^{n+1}_\xi = \frac{V_\xi}{V^{n+1}}, \tag{21}$$

$$e^{n+1}_\xi = e^n_\xi - \frac{\tau}{\rho^n_\xi} \cdot \left[g^{n+1/2}_\xi \cdot \overline{\nabla \cdot \mathbf{u}}^{n+1/2}_\xi - S^n_{\xi,xx} \overline{d}^{n+1/2}_{\xi,xx} - S^n_{\xi,yy} \overline{d}^{n+1/2}_{\xi,yy} - 2S^n_{\xi,xy} \overline{d}^{n+1/2}_{\xi,xy} + \left(S^n_{\xi,xx} + S^n_{\xi,yy}\right) \overline{d}^{n+1/2}_{\xi,\varphi}\right]. \tag{22}$$

In Eq. (22), $g^{n+1/2}_\xi = p^{n+1/2}_\xi + q^n_\xi$. The methods to calculate the materials' artificial viscosity q^n_ξ are discussed in Section 4. The bar denotes the difference counterpart of a corresponding operator (the formulas are generally known, so we skip them). In the following, we will not use the bars assuming that all the operators are difference operators. We have not described the way of updating the materials' stresses, i.e., the approximation of Eq. (16), as it is beyond the scope of this work; we only note here that these equations include components of the tensor $\mathbf{D}_\xi$.

The formulas for total pressure, viscosity, and stress deviator are the following:

$$\begin{aligned} p^{n+1/2} &= \sum_{\xi} \psi_{\xi} p_{\xi}^{n+1/2}, \\ q^{n} &= \sum_{\xi} \psi_{\xi} q_{\xi}^{n}, \\ S^{n} &= \sum_{\xi} \psi_{\xi} S_{\xi}^{n}. \end{aligned} \tag{23}$$

where the factor ψ_{ξ} is determined by the chosen closure model (see below).

Thus, the quantities that are not yet determined in Eqs. (14)–(22) include $\nabla \cdot \mathbf{u}_{\xi}$, ψ_{ξ}, q_{ξ}, and $\mathbf{D}_{\xi}$. To calculate these, one needs to use some closure relations, being the consequences of different assumptions (models) about thermodynamic states of the materials in mixed cells.

When introducing the closure relations, one should fulfill some requirements resulting from the laws of conservation.

Requirement 1 is additivity of volumes (the law of conservation of "volume")

$$\begin{aligned} V &= \sum_{\xi} V_{\xi} \\ &\text{or} \\ \sum \beta_{\xi} &= 1, \end{aligned} \tag{24}$$

the consequence of which is the relation

$$\begin{aligned} \Delta V &= \sum_{\xi} \Delta V_{\xi}, \\ &\text{or} \\ \sum \beta_{\xi} \nabla \cdot \mathrm{u}_{\xi} &= \nabla \cdot \mathrm{u}. \end{aligned} \tag{25}$$

The natural extension of relation (25) is $\mathbf{D} = \sum \beta_{\xi} \mathbf{D}_{\xi}$, which is fulfilled at

$$\mathbf{D}_{\xi} = \mathbf{D}\left(\frac{\nabla \cdot \mathbf{u}_{\xi}}{\nabla \cdot \mathbf{u}}\right). \tag{26}$$

Formula (26) is used to determine $\mathbf{D}_{\xi}$ in Eq. (22) and when approximating Eq. (16).

Requiment 2 is additivity of energies (the law of conservation of energy)

$$e = \sum \alpha_{\xi} e_{\xi}, \tag{27}$$

where the mass fraction $\alpha_{\xi} = \frac{M_{\xi}}{M}$ is given by the following expression:

$$\alpha_\xi = \beta_\xi \cdot \frac{\rho_\xi}{\rho}, \quad \text{where} \quad \rho = \sum \beta_\xi \rho_\xi. \tag{28}$$

The requirement (27) can also be written for increments of specific energies

$$\Delta e = \sum \alpha_\xi \Delta e_\xi, \tag{29}$$

where Δe_ξ is the energy increment for the constituent material, and Δe, for the whole cell.

Let us consider closure methods for the case of approximation of gas dynamics equations. In EGAK, the difference approximation of the energy equation (22) for the constituent materials has the following form:

$$\Delta e_\xi = -\frac{\tau g_\xi^{n+1/2}}{\rho_\xi^n} \nabla \cdot \mathbf{u}_\xi^{n+1/2}. \tag{30}$$

We insert their expression (30) into Eq. (29) for $\Delta \mathbf{e}_\xi$ and, using Eq. (28), obtain

$$-\frac{\tau g^{n+1/2} \nabla \cdot \mathbf{u}^{n+1/2}}{\rho} = -\frac{\tau \sum \beta_\xi g_\xi^{n+1/2} \nabla \cdot \mathbf{u}_\xi^{n+1/2}}{\rho} + \sum \alpha_\xi \Delta e_\xi^{/}. \tag{31}$$

Using Eq. (23) and the given ways of finding $\nabla \cdot \mathbf{u}_\xi$ from Eq. (31) one can obtain the values of $\Delta \mathbf{e}_\xi^{/}$ that represent additional changes in the materials' internal energy to meet the energy balance requirement.

Given that $g_\xi^{n+1/2} = g^{n+1/2}$, it follows from Eq. (31) that $\sum \alpha_\xi \Delta e_\xi^{/} = 0$. Thus, this term represents the change in the materials' internal energy as a result of their pressure relaxation. If no pressure relaxation is used, i.e., $\Delta e_\xi^{/} = 0$, then the definitions of ψ_ξ follow directly from the closing conditions.

In the general case of nonequal pressures, the requirement Eq. (31) can be fulfilled, when some conditions imposed on the function ψ_ξ are satisfied. These conditions are discussed in Section 3.

Let us write the expression for ψ_ξ as

$$\psi_\xi = \beta_\xi \lambda_\xi, \tag{32}$$

where the quantity λ_ξ is determined by the chosen model of distributing the total divergence of the mixed cell to the constituent materials from the relation

$$\nabla \cdot \mathbf{u}_\xi = \lambda_\xi \nabla \cdot \mathbf{u}. \tag{33}$$

3. Closure methods for gas dynamics equations

There is no single closure method for gas dynamics equations in mixed cells that might be suitable for all types of flow. Closing relations are quite numerous, but many of them are not used nowadays and are of historical interest only. Next, we consider the most frequently used closure methods, most of which have been implemented in EGAK.

3.1. Isotropic single-stage closure methods

It will be convenient to introduce some closure methods if we consider the 1D problem, as shown in **Figure 1**. The $i - 1/2$th cell is mixed; it contains two materials; the interface is denoted by point A. Depending on the way of velocity definition at the point A, one can derive one or another method for calculating divergences (and densities) of the materials in the mixed cell.

Figure 1. Computational mesh. Point A is the interface.

3.1.1. Method based on the equal pressure model

Method 1 uses the assumption that the materials have equal pressures (as proposed in [1]); in addition, artificial viscosities are assumed to be equal, too:

$$\begin{aligned} p_\xi &= p, \\ q_\xi &= q. \end{aligned} \tag{34}$$

For EGAK, method 1 based on the model (Eq. (34)) has been developed in [37].

This method first calculates the energy increment for the cell as a whole:

$$\begin{aligned} \Delta E &= (e^{n+1/2} - e^n)M = -\frac{\tau(p^{n+1/2} + q^n)}{\rho^n} \nabla \cdot \mathbf{u}^{n+1/2} M \\ &= -\tau(p^{n+1/2} + q^n)\nabla \cdot \mathbf{u}^{n+1/2} V^n = (p^{n+1/2} + q^n)\mu \Delta V, \end{aligned} \tag{35}$$

where $\mu = -\nabla \cdot \mathbf{u}^{n+1/2} V^n \tau / \Delta V$.

By analogy with Eq. (35), the energy increment equation for the materials can be rewritten by adding respective material indices in the quantities ΔV and ΔE. Then, using expressions (25) and (34), one can write the following closed system of equations:

$$\begin{aligned} \Delta E_\xi &= (p^{n+1/2} + q^n)\mu \Delta V_\xi, \\ \Delta V &= \sum_\xi \Delta V_\xi, \\ P\left(\frac{M_\xi}{V_\xi^n + \Delta V_\xi}, \frac{E_\xi + \Delta E_\xi}{M_\xi}\right) &= p^{n+1/2}. \end{aligned} \tag{36}$$

The system (Eq. (36)) contains $2N + 1$ equation in $2N + 1$ unknown ΔV_ξ, ΔE_ξ, and $p^{n+1/2}$ and can be solved iteratively. Solving the system of equations gives updated values of specific energies and volume fractions of the materials:

$$e_\xi^{n+1/2} = e_\xi + \frac{\Delta E_\xi}{M_\xi}, \quad \beta_\xi^{n+1/2} = \frac{V_\xi + \Delta V_\xi}{V + \Delta V}. \tag{37}$$

Among the drawbacks of this method one should note its expensiveness because of the iterative methods that are needed for solving the system (Eq. (36)) with complex equations of state. Also note that the assumption about equal pressures can turn out to be inconsistent, for example, in problems associated with energy release at every timestep.

3.1.2. Tipton's method

The model underlying Tipton's method is close to the model (34). Let us consider it as applied to the difference scheme implemented in [22] and being slightly different from Eqs. (9)–(22). At the first half-timestep, instead of Eq. (9) the method uses the equation $\mathbf{r}^{n+1/2} = \mathbf{r}^n + \tau/2 \cdot \mathbf{u}^n$, and in Eq. (13), $\chi = 1.0$. Then,

$$p_\xi^{n+1/2} = p_\xi^n - \rho_\xi^n \cdot (c_\xi^n)^2 \cdot \tau \cdot \nabla \cdot \mathbf{u}_\xi^n. \tag{38}$$

This method assumes that the pressures of the constituent materials at the half-timestep should equal the average pressure $p^{n+1/2}$, which requires

$$p^{n+1/2} = p_\xi^{n+1/2} + R_\xi^{n+1/2}, \tag{39}$$

where $R_\xi^{n+1/2}$ is different for different materials. The following equation is used to determine this quantity:

$$R_\xi^{n+1/2} = -\rho_\xi^n \cdot c_\xi^n \cdot h^n \cdot \overline{\nabla \cdot \mathbf{u}}_\xi^n, \tag{40}$$

where h^n is the characteristic mesh spacing.

Eqs. (38) and (39) can be combined into one equation.

$$p^{n+1/2} = p_\xi^n - \tilde{B}_\xi^n \nabla \cdot \mathbf{u}_\xi^n, \tag{41}$$

where $\tilde{B}_\xi^n \equiv (c_\xi^n)^2 \rho_\xi^n \cdot \tau \left[1 + \frac{h^n}{(\tau c_\xi^n)}\right]$.

Eq. (41) together with the requirement (Eq. (25)) constitutes a closed set of algebraic equations with the unknowns $\nabla\cdot\mathbf{u}_\xi^n$ and $p^{n+1/2}$. As mentioned in Section 3.1.1, the system can be solved; for $\xi = 1, 2$, the solution is the following:

$$\nabla\cdot\mathbf{u}_\xi^n = \frac{(p_\xi^n - \overline{p}^n) + \overline{B}^n\cdot\nabla\cdot\mathbf{u}^n}{\tilde{B}^n{}_\xi},$$
$$p^{n+1/2} = \overline{p}^n - \overline{B}^n\nabla\cdot\mathbf{u}^n, \tag{42}$$

where the barred terms denote average values of the quantities

$$\overline{p}^n \equiv \frac{\sum(\beta_\xi^n p_\xi^n/\tilde{B}^n_\xi)}{\sum(\beta_\xi^n/\tilde{B}^n_\xi)} \text{ and } \overline{B}^n \equiv \left[\sum\left(\frac{\beta_\xi^n}{\tilde{B}_\xi}\right)\right]^{-1}. \tag{43}$$

Eq. (42) leads to an expression for the change in the volume fractions

$$\Delta\beta_\xi^{n+1/2} = \beta_\xi^n\left(\frac{p_\xi^n - \overline{p}^n}{\tilde{B}_\xi^n}\right) + \beta_\xi^n\left[\tau\left(\frac{\overline{B}^n}{\tilde{B}_\xi^n} - 1\right)\right]\nabla\cdot\mathbf{u}^n. \tag{44}$$

To update the quantities at t^{n+1}, another assumption is made that increments of the quantities after a full timestep are twice the half timestep increments:

$$\Delta\beta_\xi^{n+1} = 2\Delta\beta_\xi^{n+1/2}$$
$$\text{and} \tag{45}$$
$$\beta_\xi^{n+1} = \beta_\xi^n + 2\Delta\beta_\xi^{n+1}.$$

The difference energy equation for the materials in the cell is the following:

$$e_\xi^{n+1} = e_\xi^n - p^{n+1/2}\frac{\Delta V_\xi^{n+1}}{M_\xi}, \tag{46}$$

where $\Delta V_\xi^{n+1} = \Delta\beta_\xi^{n+1/2}\cdot V^{n+1}$.

3.1.3. Delov's method

Method 3 based on the acoustic Riemann solver is proposed in [14]. Let us consider this method as applied to the one-dimensional problem (see **Figure 1**) with a single velocity component u. We consider the following acoustic Riemann problem in a mixed cell:

$$\begin{aligned} p = p_1^n, \quad u = u_{i-1}^{n+1/2} \quad &\text{for} \quad m < m_A, \\ p = p_2^n, \quad u = u_i^{n+1/2} \quad &\text{for} \quad m > m_A, \end{aligned} \tag{47}$$

where m is the mass variable (m_A is the cell mass).

The set of equations for the quantities similar to the Riemann invariants along the advanced and retarded characteristics has the following form:

$$u_A^{n+1/2}+\frac{p_A^{n+1/2}}{(\rho c)_1^n}=u_{i-1}^{n+1/2}+\frac{p_1^n}{(\rho c)_1^n},$$
$$u_A^{n+1/2}-\frac{p_A^{n+1/2}}{(\rho c)_2^n}=u_i^{n+1/2}-\frac{p_2^n}{(\rho c)_2^n}. \tag{48}$$

The solution to this set will be the following expression for the velocity $u_A^{n+1/2}$:

$$u_A^{n+1/2}=\frac{u_{i-1}^{n+1/2}\cdot(\rho c)_1^n+u_i^{n+1/2}\cdot(\rho c)_2^n+p_1^n-p_2^n}{(\rho c)_1^n+(\rho c)_2^n}. \tag{49}$$

Now, let us write the equation of continuity for the problem at issue. By replacing the divergence with a corresponding expression, we obtain the following equation:

$$\frac{1}{\rho^{n+1/2}}-\frac{1}{\rho^n}=\tau\cdot\frac{u_i^{n+1/2}-u_{i-1}^{n+1/2}}{M}, \tag{50}$$

where $M=\rho h$ is the linear cell mass.

A similar equation for the materials is obtained if one of the velocities is replaced with a velocity at the point A and a respective index is attached to density and mass. After substituting the expressions for velocity (49) into the equation of continuity (50), we obtain

$$\frac{1}{\rho^{n+1/2}_{\left\{\substack{1\\2}\right\}}}-\frac{1}{\rho^{n}_{\left\{\substack{1\\2}\right\}}}=\frac{\tau}{M_{\left\{\substack{2\\1}\right\}}}\cdot\left[\frac{(\rho c)^n_{\left\{\substack{2\\1}\right\}}}{(\rho c)_1^n+(\rho c)_2^n}\cdot\left(u_i^{n+1/2}-u_{i-1}^{n+1/2}\right)+\frac{p_1^n-p_2^n}{(\rho c)_1^n+(\rho c)_2^n}\right]. \tag{51}$$

Thus, as distinct from models 1–3, the present model does not employ any equilibration algorithm for pressure relaxation in this method, because the volume changes of the materials are also controlled by their pressures.

The extension of Eq. (51) in the multimaterial case without constraint of the number of materials can be written as follows:

$$\frac{1}{\rho_\xi^{n+1/2}}=\frac{1}{\rho_\xi^n}+\frac{\lambda_\xi^n}{\alpha_\xi^n}\cdot\left(\frac{1}{\rho^{n+1/2}}-\frac{1}{\rho^n}\right)+\frac{\omega\cdot\tau}{\beta_\xi^n\cdot\rho_\xi^n\cdot h^n}\left(p_\xi^n-p_\Sigma^n\right)\cdot\frac{1}{(\rho c)_\Sigma^n}, \tag{52}$$

where

$$p_\Sigma^n=\frac{1}{N}\sum_{\xi=1}^N p_\xi^n, \tag{53}$$

$$(\rho c)_\Sigma^n=\frac{1}{N}\sum_{\xi=1}^N(\rho c)_\xi^n, \tag{54}$$

$$\lambda_{\xi}^{n} = \frac{1}{(N-1)} \left(1 - \frac{(\rho c)_{\xi}^{n}}{\sum_{\zeta=1}^{N} (\rho c)_{\zeta}^{n}} \right), \tag{55}$$

where h^n is the characteristic mesh spacing, and $\omega \sim 1$ is the factor introduced to improve stability conditions of the difference scheme.

From Eq. (55), using Eq. (12), one can obtain an expression for divergences:

$$\nabla \cdot \mathbf{u}_{\xi}^{n} = \frac{\lambda_{\xi}^{n}}{\beta_{\xi}^{n}} \cdot \nabla \cdot \mathbf{u}^{n} + \omega \cdot \frac{(p_{\xi}^{n} - p_{\Sigma}^{n})}{\beta_{\xi}^{n} \cdot h^{n} \cdot (\rho c)_{\Sigma}^{n}}. \tag{56}$$

Now, let us consider the quantity λ_{ξ}^{n}. The change in the materials' volume, as we see from Eq. (55), can be written as

$$V_{\xi}^{n+1} - V_{\xi}^{n} = \frac{\lambda_{\xi}^{n}}{\beta_{\xi}^{n}} \cdot (V^{n+1} - V^{n}) + \omega \cdot \tau \cdot \frac{(p_{\xi}^{n} - p_{\Sigma}^{n})}{(\rho c)_{\Sigma}^{n}}. \tag{57}$$

From this, using the requirement (Eq. (25)), we obtain that the equality $\sum_{\xi=1}^{N} \lambda_{\xi}^{n} (\beta_{\xi}^{n})^{-1} = 1$ must be fulfilled. One can easily see that in the 1D case, for $\omega = 1$ and $N = 2$, Eq. (52) includes Eq. (51).

Now, let us show that the quantity $\lambda_{\xi}^{n} (\beta_{\xi}^{n})^{-1}$ can be taken as a function ψ_{ξ} ($\psi_{\xi} = \lambda_{\xi}^{n} (\beta_{\xi}^{n})^{-1}$) for determining the average pressure using Eq. (23). For this purpose, it will suffice to consider the case when the materials have equal pressures. Indeed, in this case, the second term in (56) is equal to zero, and to meet the energy additivity requirement (31), it will be sufficient to assume that in Eq. (23)

$$\psi_{\xi} = \lambda_{\xi}^{n} / \beta_{\xi}^{n}. \tag{58}$$

If the materials have different pressures, when we use Eq. (58), the right-hand member of the energy equation should be corrected by $\Delta e'_{\xi}$, for example, in the form of

$$\Delta e'_{\xi} = \frac{\omega \cdot \tau}{\alpha_{\xi} \rho^{n} \cdot h^{n}} \cdot \frac{\psi_{\xi}}{(\rho c)_{\Sigma}^{n}} \cdot \sum_{\xi} p_{\xi}^{n} \cdot (p_{\xi}^{n} - p_{\Sigma}^{n}). \tag{59}$$

This method provides good results in Lagrangian calculations, when the materials' volume fractions in the cell are invariable and close to each other, i.e., at $1 \gg \beta_{\xi} \gg 0$. However, the values of β in ALE calculations can range freely within $0 < \beta < 1$ and, at β close to zero, the method gives a noticeable error associated with the presence of division by β in Eq. (58). This situation is physically attributed to the fact that, in Riemann problem calculations, waves travel some distance that must be equal to or less than the size of the region occupied by each of the materials. As the choice of the timestep is based on the Courant constraint, the necessary requirement can be violated, which leads to unphysical results (e.g., a negative updated value of material volume). To fix this, additional constraints on volume increments are required.

3.1.4. Method 5 based on the equal compressibility model

Method 5 rests upon the most frequently used model of equal compressibility of materials, or to put it differently, of their equal divergences. The model has been proposed in [17]; it has also been considered in [38, 39] and is formulated as follows:

$$\nabla \cdot \mathbf{u}_\xi^n = \nabla \cdot \mathbf{u}^n. \tag{60}$$

Naturally, it is also assumed that $\nabla \cdot \mathbf{u}_\xi^{n+1} = \nabla \cdot \mathbf{u}^{n+1}$.

This method, which appears at first glance to be strange, results from the assumption that the velocity at the point A (**Figure 1**) is determined by distance-linear interpolation between nodal velocities. In the 2D case, this method is generalized on the assumption of volume-linear interpolation.

In this method, $\lambda_\xi = 1$ and formula (25) gives

$$\psi_\xi = \beta_\xi, \tag{61}$$

which ensures the fulfillment of the requirement (31) at $\Delta e'_\xi = 0$. As a result, formula (23) transforms to

$$p = \sum \beta_\xi p_\xi. \tag{62}$$

Thus, all the quantities we need to solve Eqs. (9)–(22) are defined.

Formula (62), which is natural for a homogeneous mixture of ideal gases, has a simple interpretation for a heterogeneous mixture. Let us consider a mixed cell containing two materials of volume V_0 (see **Figure 2**).

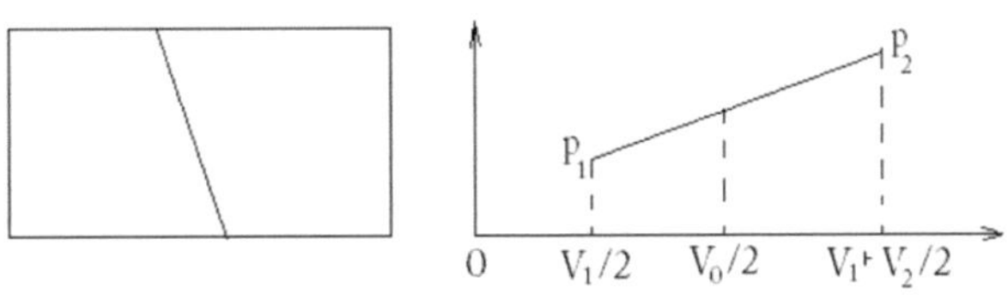

Figure 2. Graphic illustration of Formula (62).

Let us assign p_ξ to the centers of their volumes. If the volume is used as a variable, then the linear interpolation of pressure over the cell volume can be written as

$$p(V) = 2 \cdot \frac{p_2 - p_1}{V_o} \left(V - \frac{V_1}{2} \right) + p_1. \tag{63}$$

If we insert the value of $V = V_0/2$ into Eq. (63), we will obtain the very same Formula (62). It is easy to demonstrate that for the case of an arbitrary number of materials, the value of pressure at the point $V_0/2$ will also be defined by Formula (62) (first, two materials are considered, and then other materials are added one by one). Thus, in the heterogeneous case, for linear interpolation of pressure between the materials' pressures, Formula (62) defines the pressure at the volume center of the mixed cell. As the approximation of the momentum equation uses cell-related pressures, Formula (62) is consistent enough for determining the average pressure in the mixed cell.

Strengths and weaknesses of method 5 are evident. It is easy to implement and inexpensive in operation, but it can lead to nonphysical states of the materials. The point is that the materials in the mixed cell in calculations by this method are compressed uniformly, which leads to different pressures of the materials, which do not relax with time (see the test calculations in Section 5). Nevertheless, the method delivers quite acceptable results when used for flows with distinct interfaces.

3.1.5. Method 6 based on the equal pressure increments model

Bondarenko and Yanilkin [18] proposed a closure method based on the equality of pressure increments of the materials in the mixed cell. This model is mathematically expressed as

$$\rho_\xi c_\xi^2 \nabla \cdot \mathbf{u}_\xi = \rho_\zeta c_\zeta^2 \nabla \cdot \mathbf{u}_\zeta. \tag{64}$$

The condition (64) is derived as follows. For adiabatic flows,

$$\frac{\partial p}{\partial t} = \frac{\partial p}{\partial \rho} \cdot \frac{\partial \rho}{\partial t} = c^2 \rho \nabla \cdot \mathbf{u}. \tag{65}$$

Going over in Eq. (65) to the materials and claiming that $\frac{\partial p_\xi}{\partial t} = \frac{\partial p_\zeta}{\partial t}$, we obtain the condition (64).

The set of algebraic equations (64), Eq. (25) is closed, and its solution is given by

$$\nabla \cdot \mathbf{u}_\xi^n = \lambda_\xi \nabla \cdot \mathbf{u}^n, \tag{66}$$

where

$$\lambda_\xi = \left(\frac{\beta_k}{\rho_k c_k^2} \right)^{-1} \cdot (\rho_\xi c_\xi^2)^{-1}. \tag{67}$$

When this model is used in calculations for adiabatic flows, pressures will stay approximately equal, if the materials' initial pressures in the cell are equal. However, in some cases, pressures may turn out to be different: if energy release is specified for one of the materials; behind a strong shock in the mixed cell, because the condition of equal pressure increments is incorporated in the adiabatic approximation; after calculating convective fluxes at the Eulerian stage, etc. In this case, the use of this model in its original form is associated with some problems.

One of them is the following. Let $p_\xi^n \neq p_\varsigma^n$. For an ideal gas, the following estimate is true:

$$\nabla \cdot \mathrm{u}_\xi \sim \left(\rho_\xi c_\xi^2\right)^{-1} \sim \left(\gamma_\xi p_\xi\right)^{-1}. \tag{68}$$

It follows from Eq. (68) that at close values of γ_ξ, the values of $\nabla \cdot \mathbf{u}_\xi$ are inversely proportional to p_ξ. As a result, the material with a lower pressure relaxes more actively on relief, with an opposite trend in compression. However, according to the physics of the process, pressure relaxation should take place in any case. The predicted pressure for the lower pressure material can even turn out to become negative. To fix this flaw, in the case of cell expansion, method 5 uses a requirement that relative rather than absolute pressure increments of the materials should be equal. Let us write the modified equation (Eq. (65)) as

$$\nabla \cdot \mathbf{u}_\xi \approx -\frac{\Delta p_\xi}{p_\xi} \cdot \frac{p_\xi}{\tau \rho_\xi \mathrm{d}p_\xi / \mathrm{d}\rho_\xi}, \tag{69}$$

require that the condition $\Delta p_\xi / p_\xi = \Delta p_\zeta / p_\zeta$ is fulfilled, and obtain formula (70):

$$\lambda_\xi = -\frac{p_\xi}{\rho_\xi c_\xi^2} \cdot \frac{1}{\sum \beta_\mathrm{k} p_\mathrm{k} / \left(\rho_\mathrm{k} c_\mathrm{k}^2\right)}. \tag{70}$$

This value of λ_ξ will also be used in Eq. (32) to determine ψ_ξ, which meets the requirement (31) at $\Delta e'_\xi = 0$.

3.1.6. *Method 7 based on the equal velocity increments model*

This model has been proposed in [19]. The assumption that the materials' velocity increments are equal that is the consequence of the fact that the set of gas dynamics equations is solved in the one-velocity approximation. Since the materials' velocities in the mixed cell are equal at any time, it is natural that the changes in the materials' velocities at every timestep will also be equal. One can treat changes in physical quantities over a timestep as those resulting from the propagation of some perturbations. If one assumes that these perturbations are plane acoustic waves, for which $\Delta \rho / \rho = \Delta u / c$, then for $\nabla \cdot \mathbf{u}_\xi$ one can write the following expression:

$$\nabla \cdot \mathbf{u}_\xi \approx -\frac{\Delta \rho_\xi}{\rho_\xi \tau} = -\frac{\Delta u_\xi}{c_\xi \tau}. \tag{71}$$

Considering that the materials' velocity increments Δu_ξ in the mixed cell are assumed to be equal, for $\nabla \cdot \mathbf{u}_\xi$ we obtain the following relation (72):

$$c_\xi \nabla \cdot \mathbf{u}_\xi = c_\zeta \nabla \cdot \mathbf{u}_\zeta. \tag{72}$$

The set of algebraic equations (72) and (25) is closed, and its solution can be given by

http://dx.doi.org/10.5772/66858

$$\nabla \cdot \mathbf{u}_\xi^n = \lambda_\xi \nabla \cdot \mathbf{u}^n, \tag{73}$$

where

$$\lambda_\xi = \left(c_\xi \sum_k \frac{\beta_k}{c_k} \right)^{-1} = \frac{c}{c_\xi}, \tag{74}$$

with

$$c = \left(\sum_k \frac{\beta_k}{c_k} \right)^{-1}.$$

This value of λ_ξ will also be used in Eq. (32) to determine ψ_ξ, which satisfies the requirement (31) at $\Delta e'_\xi = 0$.

3.1.7. Pressure relaxation methods

The use of models 5–7 as single-stage methods in real simulations is associated with a number of problems that sometimes lead to inconsistent results. All these cases are related to the absence of the pressure relaxation mechanism for materials in mixed cells. The analysis shows that, even in model 6, despite the equal pressure increments of the materials at a timestep, equality of pressures at t^{n+1} is not always the case. The situation in the other two models is even worse.

For this reason, methods 5–7 that can be used as single-stage methods are combined with subcell pressure relaxation. Next, we consider two known relaxation methods. In all the two-stage isotropic closure models, cell divergence is redistributed among the materials only if it is different from zero. As for the second (subcell) stage of the models, it involves interactions between the materials if these are in the nonequilibrium state with no mandatory requirement that the divergence should be nonzero.

3.1.7.1. The PR method

Let us consider the PR pressure relaxation method proposed in [20]. As the materials occupy finite volumes in the mixed cell, equilibration of the materials' pressures occurs not instantaneously (instantaneous pressure equilibration occurs only at the points of the surface, along which the materials contact each other), but over some time in several timesteps.

This method calculates $\nabla \cdot \mathbf{u}_\xi$ at the timestep in two stages:

$$\nabla \cdot \mathbf{u}_\xi = \nabla \cdot \mathbf{u}_{\xi 1} + \nabla \cdot \mathbf{u}_{\xi 2}. \tag{75}$$

In Eq. (75), $\nabla \cdot \mathbf{u}_{\xi 1}$ is the material's divergence at the first stage obtained by one of the above methods. The second stage includes pressure relaxation of the materials. The second stage imposes the requirement that both $\nabla \cdot \mathbf{u}$ and the total internal energy, i.e., $\Delta E_2 = 0$, should remain unchanged at that stage. Pressure relaxation is implemented by calculating additional divergences of the materials $\nabla \cdot \mathbf{u}_{\xi 2}$ by the formulas

$$\nabla \cdot \mathbf{u}_{\xi 2} = -\frac{\Delta p_\xi}{\tau \rho_\xi c_\xi^2}, \tag{76}$$

where

$$\Delta p_\xi = A\frac{c\tau}{h}(p - p_\xi), \tag{77}$$

where p is the average pressure. Expression (76) was derived using a known relation in the adiabatic approximation, $\Delta p = -\rho c^2 \tau \nabla \cdot \mathbf{u}$. The factor $c\tau/h$, equal to the ratio of the timestep to the characteristic pressure equilibration time of the mixed cell h/c, determines the fraction of the materials' pressure difference, by which the materials' pressure equalizes. According to the meaning of expression (77), $A \sim 1$; in this case, the materials' pressures will not relax over a single timestep.

As p in the equilibration algorithm (not to confuse with the average pressure at the basic stage done by Eq. (23)) we take Eq. (62). Then, the requirement that the cell volume should be constant at this stage, $\sum \beta_\xi \nabla \cdot \mathbf{u}_{\xi 2} = 0$, will be satisfied automatically. The choice of the formula for p is ambiguous. For example, the choice based on Eq. (23) may prove to be unbeneficial. Let us illustrate this with the following example. Suppose a mixed cell contains two ideal gases having the same EOS $\gamma_1 = \gamma_2 = \gamma$, but disparate pressures and volume fractions. Let $p_1 = 1000$, $\beta_1 = 0.9$ and $p_2 = 1$, $\beta_2 = 0.1$. Using Eqs. (23) and (25) combined with the relation $\rho_\xi c_\xi^2 = \gamma_\xi p_\xi$, one can easily calculate that $p = 10$. Thus, the cell-average pressure calculated by Eq. (23) is a hundred times lower than the pressure in the material occupying nearly the whole cell. Formula (62) is free of this flaw and, as shown above, has a certain mathematical basis.

This method results in the internal energy exchange between the materials. Indeed, let us represent the total change in the materials' internal energies as

$$\Delta E = -P_+ \Delta V_+ - P_- \Delta V_-, \tag{78}$$

where $$P_+ = \sum \beta_\xi p_\xi \quad \text{if } p_\xi > p; \qquad P_- = \sum \beta_\xi p_\xi \quad \text{if } p_\xi \leq p$$ and ΔV_+ and ΔV_- are the volume changes of these materials.

With pressure equilibration, the materials with P_+ expand, so $\Delta V_+ > 0$ and $\Delta V_- < 0$. As far as it follows from the volume conservation condition that $\Delta V_+ = |\Delta V_-|$, and by definition $P_+ > P_-$, ΔE in the pressure equilibration procedure by Eq. (78) will always be negative. This situation is associated with the fact that motion of interfaces causes internal (subcell) motion in the cell, and part of the cell's internal energy converts into the subcell kinetic energy. Since the subcell kinetic energy is not taken into account in the calculations, it is returned to the materials in the form of internal energy increments $\Delta e'_\xi$ in accordance with the equation

$$-\frac{\tau}{\rho^n}\sum_{\xi}\beta_{\xi}(p_{\xi}^{n+1/2}+q_{\xi}^{n})\nabla\cdot\mathbf{u}_{\xi 2}^{\mathbf{n}+1/2}+\sum_{\xi}\alpha_{\xi}\Delta e'_{\xi}=0, \tag{79}$$

The quantity $\nabla\cdot\mathbf{u}_{\xi 2}^{\mathbf{n}+1}$ present in this expression is calculated by the formula

$$\nabla\cdot\mathbf{u}_{\xi 2}^{n+1}=\nabla\cdot\mathbf{u}_{\xi 2}^{n}(1-Ac\tau/h).$$

It remains to decide how to distribute the dissipated kinetic energy among the materials (formula (79) defines only the total dissipated energy $\Delta E=\sum_{\xi}\alpha_{\xi}\Delta e'_{\xi}$). In [20], it is assumed that $\Delta e'_{\xi}=\Delta e'$. In this case, for all the materials in the mixed cell we obtain from Eq. (79) that

$$\Delta e'_{\xi}=\Delta e'=\frac{\tau}{\rho^n}\sum_{\xi}\beta_{\xi}(p_{\xi}^{n+1/2}+q_{\xi}^{n})\nabla\cdot\mathbf{u}_{\xi 2}^{\mathbf{n}+1/2}. \tag{80}$$

Note that this pressure relaxation approach is universal, i.e., it is independent of the way, the total velocity divergence in the mixed cell is distributed among the materials. In EGAK, it is employed for the three above-mentioned methods. However, its application to physically inconsistent generation of pressure difference between the materials through the basic closure relation may lead to excessive energy exchange between them, so for each specific method its consistency needs to be validated by test simulations.

3.1.7.2. Method of Barlow, Hill and Shashkov (BHS)

This method is described in detail in [25]; here we briefly summarize and illustrate the basic concept of the method for the case of two materials, which is sufficient for understanding the whole method. In its complete form, the method has been developed for the multimaterial case; for details see [25].

This method assumes that the total change in the material volume over a timestep is the sum of two terms:

$$\Delta V_{\xi}=\Delta V_{1\xi}+\Delta V_{2\xi}, \tag{81}$$

where the subscripts 1 and 2 denote the two stages of the closure model.

The first stage employs model 5, which assumes that the materials' divergences are equal and do not require the information on the contact location in the cell. The equality of divergences means that

$$\Delta V_{1\xi}=\beta_{1\xi}\Delta V. \tag{82}$$

The second (subcell) stage uses the model based on the acoustic Riemann problem (Delov's model), which calls for the reconstruction of the contact location in the mixed cell. In cell 1234, as shown in **Figure 3**, it is the segment AB.

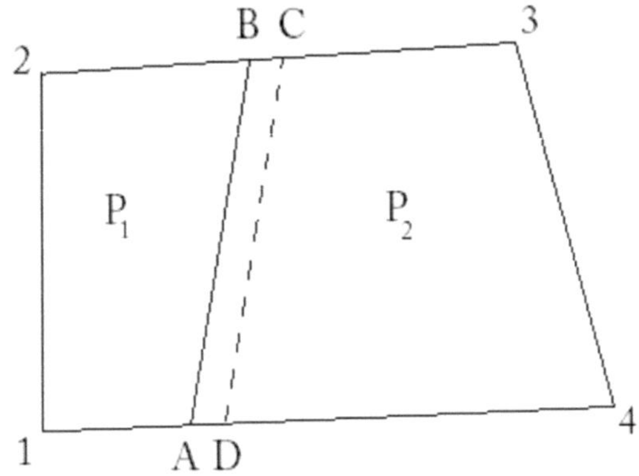

Figure 3. Contact location reconstructed after the first stage.

After the first stage, $P_1 > P_2$. Then, after the subcell stage, the contact will move to the location CD. The quantity of volume increment is represented by the quadrilateral ABCD determined based on the solution of the acoustic Riemann problem

$$\Delta V_2 = \frac{p_1 - p_2}{\rho_1 c_1 + \rho_2 c_2} S_{AB} \tau^n, \tag{83}$$

where all the unindexed quantities are taken after the first stage, and S_{AB} is the area of the boundary between the materials.

Thus,

$$\begin{aligned} \Delta V_{21} &= \Delta V_2, \\ \Delta V_{22} &= -\Delta V_2. \end{aligned} \tag{84}$$

The updated volumes of the materials calculated by formula (81) accounting for the volume increments must satisfy the following inequalities:

$$V^{n+1} > V_\xi^{n+1} > 0, \tag{85}$$

which, however, is not always the case for the same reason as in Delov's method (see the remark at the end of Section 3.1.3). Therefore in [25], the authors introduce constraints on volume increments, which complicate the method significantly, especially in the multimaterial case.

3.2. Anisotropic closure models

3.2.1. *Model fundamentals*

Let us consider two limiting cases of contact location relative to the wave motion (shock, acoustic, or elastic wave) presented in **Figure 4**, in which cells contract or expand, i.e., the divergence is nonzero. *In the first case* (**Figure 4a**), most of motion is normal to the interface, so all the above models 1–7, each having its own accuracy, are suitable. *In the second case* (**Figure 4b**) most of motion occurs along the interface, while the lateral motion is insignificant and is therefore auxiliary. It means that the materials contract or expand tangentially to the interface; thus, equality of compressibilities, i.e., model 5, may be more consistent in this case. Indeed, calculations show that, for example, using models 6 and 7 for such flows in the elastoplastic case one can obtain a considerable error, while model 5 provides good accuracy.

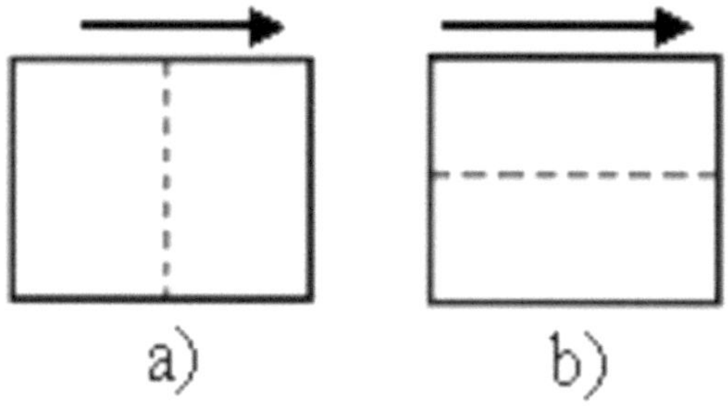

Figure 4. Two cases of contact location relative to wave motion.

The above fact implies that closure models 5–7 are inappropriate for modeling such flows. Thus, to ensure acceptable modeling accuracy for the two different types of flow (in different directions relative to the interface), different closure relations need to be used. For this purpose, two-stage models are proposed.

3.2.2. *The ACM-1 model*

At the first stage in the anisotropic model ACM-1 [34], matter in the mixed cell moves as a whole, and all nonuniformities (including the interface) are assumed to be frozen. The freezing condition in terms of closing in the first approximation means that the materials' divergences are equal.

The second stage includes relaxation of pressure (and stresses) concurrent with such motion. The work [34] suggests using the above pressure relaxation procedure at this stage with the degree of relaxation made dependent upon the mutual orientation of the velocity direction and the interface. If the velocity is normal to the interface, the pressure relaxation is the highest, and if the velocity is directed along the interface, it is the smallest.

In the implementation of the ACM-1 model, two stages are used to calculate $\nabla \cdot \mathbf{u}_\xi$ at a timestep

$$\nabla \cdot \mathbf{u}_\xi = \nabla \cdot \mathbf{u}_{\xi 1} + \nabla \cdot \mathbf{u}_{\xi 2}. \tag{86}$$

In Eq. (86), $\nabla \cdot \mathbf{u}_{\xi 1}$ is calculated at the first stage in accordance with closure model 1:

$$\nabla \cdot \mathbf{u}_{\xi 1} = \nabla \cdot \mathbf{u}. \tag{87}$$

The second stage includes pressure relaxation of the materials in mixed cells according to the algorithm, basically similar to the algorithm described in Section 3.1.7.1. The only distinguishing feature is that for the ACM-1 model, the coefficient A in Eq. (76) depends on the mutual orientation of the velocity and the interface. The total divergence is written as the sum of two components:

$$\nabla \cdot \mathbf{u} = \nabla \cdot \mathbf{u}_\tau + \nabla \cdot \mathbf{u}_n, \tag{88}$$

obtained by velocity decomposition to two components: along the interface ($\mathbf{u}_\tau$) and normal to it ($\mathbf{u}_n$). We also assume that

$$A = A_0 \cdot \frac{\nabla \cdot \mathbf{u}_n}{\nabla \cdot \mathbf{u}}, \tag{89}$$

where A_0 is some constant.

Thus, the coefficient A is a variable in this case. If the velocity is normal to the interface (**Figure 4a**), $\nabla \cdot \mathbf{u}_n = \nabla \cdot \mathbf{u}$ and $A = A_0$; this is the case with the highest pressure relaxation. If the velocity is directed along the interface (**Figure 4b**), $\nabla \cdot \mathbf{u}_n = 0$ and $A = 0$, i.e., there is no pressure relaxation at all, and only the first stage of the closure method is in effect, i.e., formula (87).

The constant A_0 in Eq. (89) must be determined based on test simulation results. In [34], its value was determined in several simulations, and it proved to coincide with the value of the constant A in Section 3.1.7.1, i.e., $A_0 = 1$.

3.2.3. *The ACM-2 model*

The anisotropic model ACM-2 is formulated as follows. We divide the divergence of the entire cell and its materials into two components: normal and tangential (relative to the contact location):

$$\begin{array}{c} \nabla \cdot \mathbf{u} = \nabla \cdot \mathbf{u}_n + \nabla \cdot \mathbf{u}_\tau \\ \nabla \cdot \mathbf{u}_\xi = \nabla \cdot \mathbf{u}_{\xi n} + \nabla \cdot \mathbf{u}_{\xi \tau} \end{array}. \tag{90}$$

Along the interface, the materials are assumed to have equal compressibilities, i.e., the distribution of the corresponding divergence component among the materials is described by the relation

$$\nabla \cdot \mathbf{u}_{\xi \tau} = \nabla \cdot \mathbf{u}_\tau. \tag{91}$$

As for the divergence in the direction normal to the interface, it can be distributed to the materials using any of closure models 5–7. In this work, we use model 7, which, as shown in [40, 41], is the most widely applicable with respect to modeling different kinds of flow. It follows from it that

$$\begin{array}{c} \nabla \cdot \mathbf{u}_{\xi n} = \lambda_\xi \nabla \cdot \mathbf{u}_n \\ \lambda_\xi = \left(c_\xi \sum_k \frac{\beta_k}{c_k} \right)^{-1} \end{array}. \tag{92}$$

Once this part of divergence is distributed to the materials, relaxation of their pressures is carried out by the algorithm described in Section 3.1.7.1, which makes an additional contribution, $\nabla \cdot \mathbf{u}'_{\xi n}$, to the divergence $\nabla \cdot \mathbf{u}_{\xi n}$:

$$\widetilde{\nabla\cdot\mathbf{u}}_{\xi n} = \nabla\cdot\mathbf{u}_{\xi n} + \nabla\cdot\mathbf{u}'_{\xi n}. \tag{93}$$

The ultimate formula for the distribution of $\nabla\cdot\mathbf{u}$ among the materials will be

$$\nabla\cdot\mathbf{u}_{\xi} = \widetilde{\nabla\cdot\mathbf{u}}_{\xi n} + \nabla\cdot\mathbf{u}_{\xi\tau}. \tag{94}$$

4. Artificial viscosity

For mixed cells, closure relations for calculating divergences of constituent materials are insufficient. Such cells require an additional relation to determine the artificial viscosity for the cell as a whole and for each individual material. Two approaches can be used to calculate the viscosity of the materials.

The first approach is based on viscosity calculations directly for the materials based on their individual parameters. The viscosity for matter as a whole is then calculated based on the procedure used to calculate the average pressure from the materials' pressures.

The second approach is based on the calculations of the viscosity for the cell as a whole based on the cell-average parameters of matter and its distribution to the materials according to some assumptions on the way of its distribution.

The first approach is more expensive both in terms of research effort, and in terms of computations; therefore, the less complicated second approach has been developed more widely. This work mostly considers viscosity definition procedures based on the second approach.

4.1. Artificial viscosity for matter as a whole

The viscosity of matter as a whole in EGAK is calculated in the cells, and it is a combination of the von Neumann and Richtmyer type quadratic viscosity and linear viscosity

$$\begin{cases} q^n = C_1\cdot\rho^n(h^n\nabla\cdot\mathbf{u}^n)^2 + C_0\cdot\rho^n\cdot c\cdot h^n\nabla\cdot\mathbf{u}^n & \text{if } \nabla\cdot\mathbf{u}^n < 0 \\ q^n = 0 & \text{if } \nabla\cdot\mathbf{u}^n \geq 0 \end{cases}. \tag{95}$$

In Eq. (95), $C_1 = 1$ and $C_0 = 0.2$ are the fixed coefficients. In addition, expression (95) contains the characteristic dimension h and the divergence $\nabla\cdot\mathbf{u}^n$ of the cell. Here, we will not address the issues of determining these quantities, as it does not matter for our purposes. Let us consider the approaches to viscosity distribution to materials.

4.2. Artificial viscosity of materials in mixed cells

The research summarized below has been carried out in [36]. Viscosity calculations for materials represent an ambiguous problem; for solving it correctly, it is usually insufficient to have data on the subcell behavior of the materials. The way of the materials' viscosity representation governs the distribution of energy dissipated in the cell on shock propagation to the materials. The problem of energy distribution to the materials is a subcell problem that lacks information needed for obtaining the exact solution. In reality, the shock front width is generally much smaller than the mesh spacing, so the shock propagates through each material in the heterogeneous mixture practically independently. Shock parameters in each material are different, and they differ from average values in the mixed cells, so it is practically impossible to determine uniquely the fraction of dissipated energy for each material. Pressures and velocities of the materials are different behind the shock, and the processes of pressure and velocity relaxation provide additional redistribution of dissipated energy between the materials.

When considering the approaches to viscosity definition for the materials, let us characterize these approaches in terms of dissipated energy distribution among the materials and resulting changes in the materials' pressures at a timestep. Before we consider different kinds of material viscosities, let us note that the cell-average value of viscosity is determined using Eq. (23), i.e., $q = \sum \psi_\xi q_\xi$, $\psi_\xi = \beta_\xi \lambda_\xi$, where λ_ξ is a normalizing factor (see Section 3), which imposes a constraint on the closure methods (all the procedures described below have been studied using EGAK on methods 5–7 and ACM-1).

In accordance with the difference scheme of EGAK, the viscosity-related change in the materials' specific internal energy over a timestep is given by the formula

$$\Delta e_\xi = -\frac{q_\xi^n}{\rho_\xi^n} \cdot \lambda_\xi^n \cdot \tau \cdot (\nabla \cdot \mathbf{u}^{n+1/2}). \tag{96}$$

The viscosity-related change in the materials' pressure at a timestep can be obtained as follows. For adiabatic flows, it holds true that

$$\Delta p \approx \left(\frac{\partial p}{\partial \rho}\right)_S \cdot \Delta \rho \approx -\rho \cdot c^2 \cdot \tau \cdot \nabla \cdot \mathbf{u}. \tag{97}$$

considering that $\Delta \rho \approx -\rho \cdot \nabla \cdot \mathbf{u} \cdot \tau$.

Using the EOS $p = P(\rho,e)$, the total pressure change at a timestep in the general case can be represented as

$$\Delta p \approx \left(\frac{\partial p}{\partial \rho}\right)_e \Delta \rho + \left(\frac{\partial p}{\partial e}\right)_\rho \Delta e. \tag{98}$$

For adiabatic flows, the energy increment is calculated by the formula

$$\Delta e = -\frac{p}{\rho} \cdot \tau \cdot \nabla \cdot \mathbf{u}. \tag{99}$$

Substituting this expression into Eq. (98) and comparing it with Eq. (96), considering Eq. (97), we obtain

$$\left(\frac{\partial p}{\partial \rho}\right)_e + \left(\frac{\partial p}{\partial e}\right)_\rho \cdot \frac{p}{\rho^2} = c^2. \tag{100}$$

On shock propagation, the energy increment is calculated by the formula

$$\Delta e = -\frac{p+q}{\rho} \cdot \tau \cdot \nabla \cdot \mathbf{u}. \tag{101}$$

Using Eq. (100) from Eq. (98), we obtain the total pressure increment in the form of

$$\Delta p \approx -\left[\rho \cdot c^2 + \left(\frac{\partial p}{\partial e}\right)_\rho \cdot \frac{q}{\rho}\right] \tau \cdot \nabla \cdot \mathbf{u}. \tag{102}$$

It follows from Eq. (102) that the materials' viscosity-related pressure increment at a timestep equals

$$\Delta p_{q\xi} \approx -\left(\frac{\partial p_\xi}{\partial e_\xi}\right)_\rho \cdot \frac{q_\xi^n}{\rho_\xi^n} \cdot \lambda_\xi^n \cdot \tau \cdot (\nabla \cdot \mathbf{u}_\xi^{n+1/2}). \tag{103}$$

Now, let us consider models to material viscosity definition. Six models have been explored. **Table 1** provides their descriptions and formulas and changes in specific energy and pressure in accordance with Eqs. (96) and (103).

Thus, in these six models to the materials' viscosity definition, distribution of dissipated energy to the materials is differently dependent on their density, speed of sound, and divergence. The choice of one model or another depends both on the closure method, and on the modeled problem. Based on the test calculations in [36], the best performance was demonstrated by model 3.

	Description of the approach to calculating $\mathbf{q}_\xi$	Formula	$\Delta \mathbf{e}_\xi$	$\Delta \mathbf{p}_\xi$
1	Equal to the cell-average viscosity	$q_\xi = q$	λ_ξ/ρ_ξ	$\left(\frac{\partial p_\xi}{\partial e_\xi}\right)_\rho \cdot \frac{\lambda_\xi}{\rho_\xi}$
2	Viscosity with its quantities $\rho_\xi, h_\xi = \beta_\xi h, \nabla \cdot \mathbf{u}_\xi$	$q_\xi = q \frac{\rho_\xi \beta_\xi^2 \lambda_\xi^2}{\sum \rho_k \beta_k^3 \lambda_k^3}$	$\beta_\xi^2 \lambda_\xi^3$	$\left(\frac{\partial p_\xi}{\partial e_\xi}\right)_\rho \beta_\xi^2 \xi_\xi^3$
3	Proportional to material densities $q_\xi \sim \rho_\xi$	$q_\xi = q \frac{\rho_\xi}{\sum \rho_k \beta_k \xi_k}$	λ_ξ	$\left(\frac{\partial p_\xi}{\partial e_\xi}\right)_\rho \lambda_\xi$
4	Same energy increment $\Delta e_\xi = \Delta e_\zeta$	$q_\xi = q \frac{\rho_\xi}{\rho \lambda_\xi}$	$\Delta e_\xi = \Delta e_\varsigma$	$\left(\frac{\partial p_\xi}{\partial e_\xi}\right)_\rho$
5	Same increment $\Delta p_\xi \Delta p_\xi = \Delta p_\zeta$	$q_\xi = q \frac{\rho_\xi}{\lambda_\xi \left(\frac{\partial p_\xi}{\partial e_\xi}\right)_\rho \sum \frac{\beta_k \rho_k}{(\partial p_k/\partial e_k)_\rho}}$	$\frac{1}{(\partial p_\xi/\partial e_\xi)_\rho}$	$\Delta p_\xi = \Delta p_\varsigma$
6	Proportional to Δp_ξ in adiabatic approximation $q_\xi \sim \Delta p_\xi$	$A \times \rho_\xi c_\xi^2 = (\partial p_\xi/\partial e_\xi)_\rho q_\xi/\rho_\xi$, where A is the proportionality factor	$\frac{\lambda_\xi \rho_\xi c_\xi^2}{(\partial p_\xi/\partial e_\xi)_\rho}$	$\lambda_\xi \rho_\xi c_\xi^2$

Table 1. Models to calculate the viscosity and specific energy and pressure increments of the materials.

5. Method for calculating mixed cells with vacuum

One of the materials available in EGAK is a zero-pressure "vacuum." For the case of vacuum, a special algorithm has been developed, which is the same for closure methods 1 and 5–7. The development of this algorithm was motivated by the fact that the pressure for vacuum is specified rather than controlled by the closure method.

In a mixed cell with vacuum, two cases are possible: $\nabla \cdot \mathbf{u} > 0$ and $\nabla \cdot \mathbf{u} \leq 0$.

The case of $\nabla \cdot \mathbf{u} > 0$. In this case, it is assumed that

$$\nabla \cdot \mathbf{u}_\xi = \nabla \cdot \mathbf{u}. \tag{104}$$

The case of $\nabla \cdot \mathbf{u} \leq 0$. In this case, the cell volume is assumed to decrease only due to a decrease in the vacuum volume, while the volumes of the other materials change only after the vacuum gets closed. This can be represented by the following formula:

$$\begin{aligned}
\nabla \cdot \mathbf{u}_\xi &= -\frac{2(\rho_\xi^{n+1} - \rho_\xi^n)}{(\rho_\xi^{n+1} + \rho_\xi^n)\tau}, \\
\rho_\xi^{n+1} &= \frac{V^n \beta_\xi^n \rho_\xi^n}{V^{n+1} \beta_\xi^{n+1}}, \\
\beta_\xi^{n+1} &= \frac{\beta_\xi^n (1 - \beta_{vac}^{n+1})}{(1 - \beta_{vac}^n)}, \\
\beta_{vac}^{n+1} &= \frac{\beta_{vac}^n V^n - \min\{\Delta V; \beta_{vac}^n V^n\}}{V^{n+1}}.
\end{aligned} \tag{105}$$

The following is proposed for the anisotropic methods ACM-1 and ACM-2: we represent the total divergence, like in Eq. (88), as a sum of two items, $\nabla \cdot \mathbf{u}_\tau$ and $\nabla \cdot \mathbf{u}_n$. If the cell expands, i. e., if $\nabla \cdot \mathrm{u} \geq 0$, then $\nabla \cdot \mathbf{u}_\xi = \nabla \cdot \mathbf{u}$.

When the cell contracts, i.e., when $\nabla \cdot \mathbf{u} < 0$:

- if $\nabla \cdot \mathbf{u}_n < 0$, then $\nabla \cdot \mathbf{u}_\xi = \nabla \cdot \mathbf{u}_{\xi\tau}$, $\nabla \cdot \mathbf{u}_{vac} = \nabla \cdot \mathbf{u} - \beta_\xi \cdot \nabla \cdot \mathrm{u}_\xi / \beta_{vac}$;

- if $\nabla \cdot \mathbf{u}_n \geq 0$, then $\nabla \cdot \mathbf{u}_\xi = \nabla \cdot \mathbf{u}$.

Let us consider the cases illustrated in **Figure 4**. Suppose the cell is contracting. Then, if the velocity is normal to the interface (**Figure 4a**), we have $\nabla \cdot \mathbf{u}_n = \nabla \cdot \mathbf{u}$, $\nabla \cdot \mathbf{u}_\tau = 0$, and as $\nabla \cdot \mathbf{u}_n < 0$, we obtain $\nabla \cdot \mathbf{u}_{\text{vacuum}} = \nabla \cdot \mathbf{u}$, i.e., only the vacuum is contracting. If the velocity is directed along the interface (**Figure 4b**), $\nabla \cdot \mathrm{u}_n = 0$ and $\nabla \cdot \mathbf{u}_\xi = \nabla \cdot \mathbf{u}$.

6. Test problems and results

The author does not have results of testing all of the above closure methods on the problems modeled in the section, so below he basically presents the results for methods 1

(*P*), 5($\nabla \cdot \mathbf{u}$), 6(Δp), 7(Δu), and, correspondingly, for the methods $\nabla \cdot$ u-*PR*, Δp-*PR*, Δu-PR, and ACM-1 and ACM-2, which have been developed with his direct participation. These methods have been tested on numerous problems, including those not included in this work (see [40, 41]). It does not seem possible to present results of all such calculations, so the author confines himself to three one-dimensional and one two-dimensional problem having analytical solutions. All the 1D calculations were done in Lagrangian variables, and the 2D one in Eulerian.

The following unified types of data processing are provided for all the calculations.

- Tables with quantities characterizing the order of convergence of the integral error of basic quantities in the calculations in the L_1 norm at a reference time.
- Tabulated values of basic quantities in mixed cells at a reference time.

The error is calculated by formula (106):

$$\delta y = \left\| y_{\text{comp}} - y_{\text{exact}} \right\|_1 = Ah^{\sigma}, \tag{106}$$

where h is the initial mesh spacing and y_{comp} and y_{exact} are the calculated and the exact value of the quantity at the cell center, respectively.

In EGAK calculations, mixed cells were of the same size as pure cells in pure-cell calculations. In the mixed cell calculations, domain coordinates were shifted to the right at a distance of $\delta x = h/2$, where h is the mesh spacing in the corresponding calculation. In other studies, the size of mixed cells was doubled, but their number was less than one cell.

In addition, a two-dimensional problem of elastic wave propagation in a thin plate is discussed, for which only EGAK results are presented and the analytical solution is given in [42]. For this problem, we have calculated the velocity of a longitudinal elastic wave in a plate. In [42], we have also solved this problem numerically using EGAK in the absence of mixed cells, which demonstrated good accuracy of the calculations in this setup. In [34], this problem has been solved numerically in the setup, in which interfaces are misaligned with grid lines thus producing mixed cells.

6.1. The water-air shock tube problem

Setup. The problem has the following initial conditions [43]:

$$(\gamma, p_\infty, \rho, e, p, \mathbf{u}) = \begin{cases} (4.4,\ 6 \cdot 10^8,\ 10^3,\ 1.07 \cdot 10^6,\ 10^9,\ 0), & \text{if} \quad 0 \le x \le 0.7, \\ (1.4,\ 0,\ \ 50,\ 5 \cdot 10^4,\ \ 10^6,\ 0), & \text{if}\ 0.7 < x \le 1. \end{cases} \tag{107}$$

The EOS of water is $p = (\gamma - 1)\rho e - \gamma p_\infty$, for which the squared speed of sound is calculated by the formula $c^2 = \gamma(\gamma - 1)(e - p_\infty/\rho) = \gamma(p + p_\infty)/\rho$.

The final time of the calculation is $t = 2.2 \cdot 10^{-4}$. The exact solution to the problem has been obtained in [44]. The calculations were carried out on meshes having 250, 500, and 1000 cells.

Results. Table 2 presents the values of the factor A and the order of convergence of the integral error in the basic quantities at $t = 2.2 \cdot 10^{-4}$, and **Table 3** shows the exact and calculated values of the basic quantities for the closure methods, for which data are available in publications. **Figure 5** shows the L_1 norm of the absolute error as a function of h.

	p ($\times 10^{-6}$)		ρ		e ($\times 10^{-3}$)		u	
Method	A ($\times 10^{-2}$)	σ	A ($\times 10^{-2}$)	σ	A ($\times 10^{-2}$)	σ	A ($\times 10^{-2}$)	σ
Pure (EGAK)	**5.13**	**0.84**	**1.92**	**0.86**	**1.3**	**0.84**	**6.0**	**0.91**
$\nabla \cdot \mathbf{u}$	4	0.79	12.4	1.10	14.5	1.16	3.56	0.79
p	4.43	0.80	2.5	0.89	1.14	0.81	1.8	0.71
Δp	8.2	0.83	5.08	0.93	5.18	0.96	3.62	0.79
Δu	8.29	0.84	16.2	0.96	9.41	0.90	15.1	0.92
$\nabla \cdot \mathbf{u}$-PR	4.1	0.79	2.28	0.89	1.03	0.80	1.57	0.70
Δp-PR	7.11	0.87	4.20	0.88	5.78	0.92	6.97	0.88
Δu-PR	6.26	0.85	2.49	0.83	2.04	0.83	4.27	0.84
Pure (Delov)	**4.96**	**0.97**	**2.15**	**0.94**	**1.06**	**0.87**	**3.73**	**0.91**
Delov	7.42	0.94	8.45	0.96	18.4	0.98	22.1	1.03
Pure (K&S, Tipton)	**7.65**	**0.99**	**2.18**	**0.97**	**0.7**	**0.90**	**4.12**	**0.92**
K&S	6.95	0.99	3.63	1.01	3.09	1.01	7.98	1.03
Tipton	4	0.79	12.4	1.10	14.5	1.16	3.56	0.79

Table 2. Factor A and the rate of convergence σ with mesh refinement.

Method	p_1 ($\times$ 10^{-7})	p_2 ($\times$ 10^{-7})	ρ_1 ($\times$ 10^{-2})	ρ_2 ($\times$ 10^{-2})	e_1 ($\times$ 10^{-5})	e_2 ($\times$ 10^{-5})
Exact	**1.599**	**1.599**	**8.050**	**2.204**	**9.704**	**1.813**
Pure (EGAK)	1.599	1.599	7.993	1.535	9.773	2.605
p	1.595	1.595	7.764	1.377	10.062	2.896
$\nabla \cdot \mathbf{u}$	3.111	0.078	8.030	0.401	9.784	0.484
Δp	99.034	1.351	9.972	4.606	10.707	7.332
Δu	39.664	0.065	8.931	0.174	10.001	0.931
$\nabla \cdot \mathbf{u}$-PR	1.594	1.594	7.579	1.504	10.306	2.650
Δp-PR	1.599	1.599	7.455	0.471	10.477	8.479
Δu-PR	1.599	1.599	7.395	1.104	10.562	3.620
Pure (Delov)	1.599	1.599	8.113	1.409	9.629	2.835
Delov	1.595	1.594	8.044	0.249	9.711	16.015
Pure (K&S, Tipton)	1.599	1.599	7.983	1.669	9.785	2.394
K&S	1.599	1.599	7.352	1.364	10.625	2.930
Tipton	1.599	1.599	7.315	0.591	10.680	6.765

Table 3. Exact and calculated values of the basic quantities in mixed cells on a mesh having 1000 cells at $t = 2.2 \times 10^{-4}$.

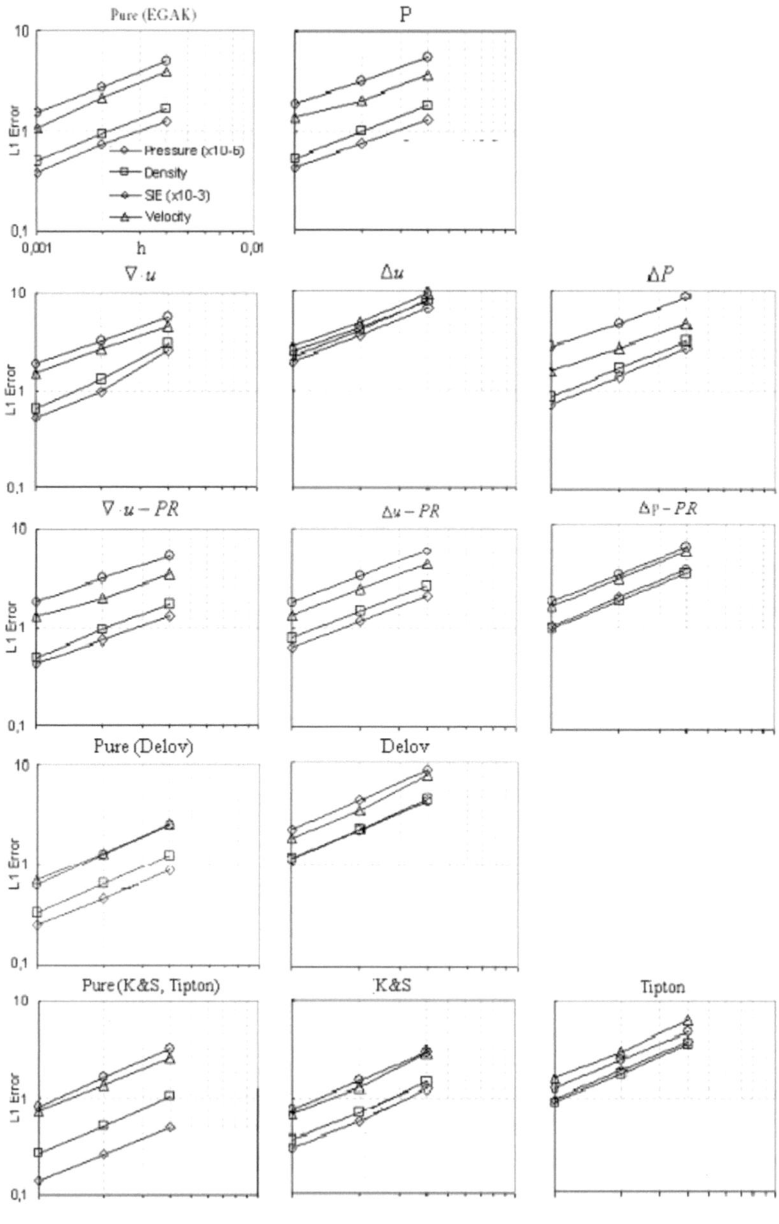

Figure 5. L_1 norm of the absolute error as a function of h.

6.2. The mixed-material shock transition problem

Setup: The problem has been proposed in [18]. The domain $-2 < x < 1$ contains a mixture of two ideal gases having the following parameters: $\rho_1^0 = 1,\ e_1^0 = 0,\ \gamma_1 = 3,\ \beta_1^0 = 0.5$ (material 1) and $\rho_2^0 = 1,\ e_2^0 = 0,\ \gamma_2 = 1.2,\ \beta_2^0 = 0.5$ (material 2). A constant velocity of $u = 2$ is given at the left boundary. Due to the specified boundary condition, a strong shock starts moving across the mixture. The problem has an analytical solution obtained in [20] assuming that the materials' pressures are equal.

The values of densities are determined based on the condition that the shock is strong for each of the materials: $\rho_\xi = (\gamma_\xi + 1)/(\gamma_\xi - 1)\rho_\xi^0$. It is implicitly supposed here that only one shock travels across the gases (additional waves reverberating between the interfaces are ignored). The volumes occupied by the materials behind the shock equal $V_\xi = V_\xi^0 \rho_\xi^0 / \rho_\xi$. The average density behind the shock front then equals

$$\rho = \frac{\sum M_\xi}{\sum V_\xi} = \frac{\sum V_\xi^0 \rho_\xi^0}{\sum V_\xi} = \frac{\sum V_\xi^0}{\sum V_\xi(\gamma_\xi - 1)/(\gamma_\xi + 1)} \rho^0. \tag{108}$$

The laws of conservation of mass, momentum, and total energy for the shock (the Rankine-Hugoniot relations) traveling across each of the materials make it possible to find the parameters of the gases behind the shock front:

$$D = \rho/(\rho - \rho^0)\mathbf{u}, \tag{109}$$

$$\mathbf{u}^2 = p(1/\rho^0 - 1/\rho), \tag{110}$$

$$e = 0.5p(1/\rho^0 - 1/\rho). \tag{111}$$

Using Eq. (109), we obtain the shock velocity, using Eq. (110), the average pressure, and using Eq. (111), the average energy of the mixture. The pressures of the materials are equal to the average pressure due to the assumption we made, and the energies of the materials can be obtained from the corresponding EOS.

This problem was calculated on a mesh having 600 cells. This problem is also of interest in terms of examining the effect of the approach to calculate the artificial viscosity for the materials.

Results: This problem differs from the two problems discussed above. First, there are no pure cells, so pure-cell calculations are inapplicable in this case. Second, only some of the above dependences can be obtained for this problem, and these are presented below. In particular, it has almost no sense to perform convergence calculations for this problem, because the steady-state solution in the mixed cells does not depend on the mesh spacing. **Table 4** presents the values of the parameters in the mixed cell with $x = 0.2$ at $t = 2$ for the materials behind the shock obtained using Eqs. (109)–(111) and in the calculations (for the materials' viscosities by approach 3). **Table 4** shows the results obtained with different viscosities for the method $\nabla \cdot$ u-PR.

6.3. 2D problem of elastic wave propagation in a plate

Next, we consider a 2D problem, in which a longitudinal elastic wave propagates in a thin plate and the wave velocity for which has been obtained analytically in [42]. The problem has been calculated using EGAK in the absence of mixed cells in [42] and with mixed cells in [34]. As a surrounding medium in the problem, we used air or vacuum.

Method	D	p_1	p_2	ρ_1	ρ_2	e_1	e_2
Exact	**2.839**	**5.677**	**5.677**	**2.0**	**11**	**1.419**	**2.581**
p	2.997	5.992	5.992	1.694	13.422	1.769	2.232
$\nabla \cdot \mathbf{u}$	3.456	13.219	0.581	2.379	2.379	2.778	1.222
Δp	2.830	5.753	5.477	2.032	10.596	1.416	2.584
Δu	2.827	5.886	5.324	2.053	10.372	1.434	2.567
$\nabla \cdot \mathbf{u}$-PR	2.859	5.715	5.715	1.956	11.253	1.461	2.539
Δp-PR	2.837	5.668	5.668	2.011	10.939	1.409	2.591
Model 1	2.916	5.820	5.820	1.844	12.013	1.578	2.422
Model 2	2.965	5.909	5.909	1.762	12.720	1.677	2.323
Model 3	2.817	5.640	5.640	2.047	10.754	1.378	2.622
Model 4	2.859	5.715	5.715	1.956	11.253	1.461	2.539
Model 5	2.535	5.049	5.049	3.503	7.697	0.720	3.279
Model 6	3.11	6.207	6.207	1.545	15.587	1.991	3.110

Table 4. Exact and calculated values of the basic quantities in the cell with $x = 0.2$ at $t = 2$ on a mesh having 600 cells (model 1 is the way of cell viscosity distribution to the materials for the closure method Δu-PR).

Setup: In the calculations, a titanium projectile of length $L = 10$ cm flying at a velocity of $v_0 =$ 0.01 km/s surrounded by air or vacuum hits a "rigid" wall. This produces an elastic wave in the projectile traveling toward its rear surface. **Figure 6** shows a schematic drawing of the initial problem geometry; $H = 1$ is the thickness of the wall. The calculations were carried out on a fixed mesh having a mesh spacing of $h = 0.2$ cm. The parameters of the EOS and the model of matter for the materials are shown in **Tables 5** and **6**.

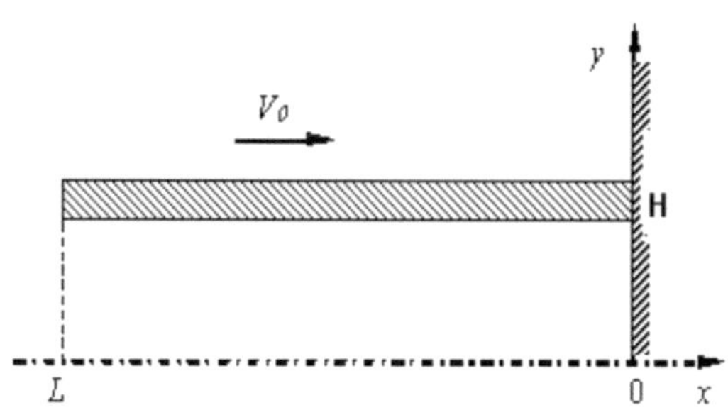

Figure 6. Geometry of the problem of elastic wave propagation in a plate.

The mesh was constructed in such a way that the projectile was initially surrounded by mixed cells containing titanium and vacuum (air) with a ratio $\beta = 0.5$. We also carried out calculations on an oblique mesh with a varied volume fraction. The field of volume fractions of titanium and a mesh fragment for this simulation are shown in **Figure 7**.

ρ_0 (g/cm^3)	c_0 (km/s)	n	Γ
4.5	4.842	3.4243	1.18

Table 5. Mie-Grüneisen EOS parameters.

	b (GPa)	k	c	m	C_v (kJ/(g × K))	T_m (K)	G (GPa)	ν
1.098	1.092	0.93	0.014	1.1	580 × 10^{-6}	1878	43	0.32

Table 6. Johnson-Cook model parameters.

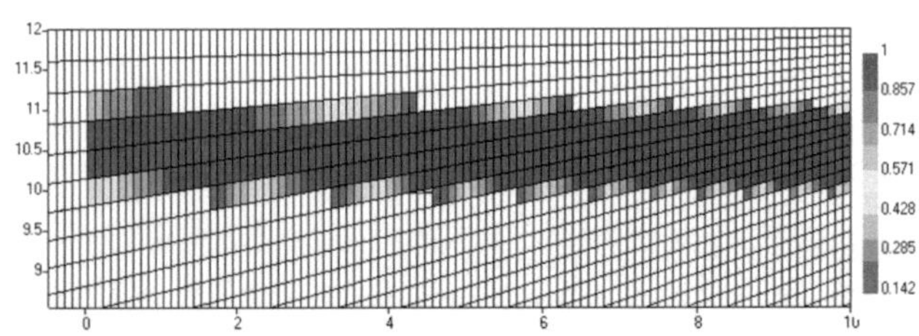

Figure 7. Distribution of volume fractions in the simulation on a fragment of the oblique mesh.

Results: In this problem, there is certain difficulty determining the longitudinal wave velocity. To address this difficulty, the following approach has been proposed in [42]. Suppose the elastic wave front does not "smear" as it propagates in the material. As the projectile hits the "rigid" wall, the velocity of the projectile material behind the wave should be zero. Therefore, the rate of deceleration of the projectile's center of mass can be related to the elastic wave velocity. **Figure 8** shows the time-history plots for the velocity of the projectile's center of mass for three simulations. The plots demonstrate that these dependences are nicely approximated by a linear function ($v = v_0 - At$). The time it takes the step-like elastic wave to travel all the way along the projectile is $T = v_0/A$. Here, T is the time, at which $v = 0$, which corresponds to the time of wave traveling all the way along the projectile. The longitudinal wave velocity is then defined as $c_w = L/A$. The error associated with the displacement of the projectile's rear end as the wave travels all the way along the projectile can be neglected, because the material's velocity is small compared to the wave velocity. **Table 7** shows the values of longitudinal wave velocities for all simulations.

The calculations of this 2D problem demonstrated that, for both of the anisotropic closure methods, the difference between the calculated elastic wave velocity and the exact solution is ~4%, whereas for the method Δu-PR it is ~10%. No comparison with other methods was made, because the Δu-PR method proved to be the most versatile among all the methods in EGAK as applied to a wide range of different problems.

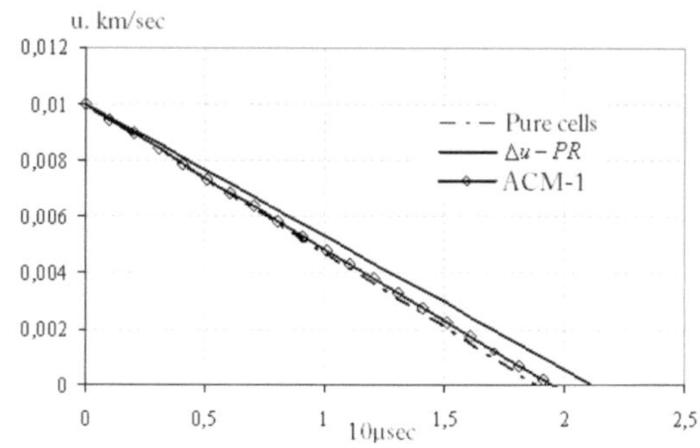

Figure 8. Velocity of the plate's center of mass as a function of time in calculations with different closure conditions for mixed cells (simulations with air).

Closure method	Surrounding medium	c_w (km/s)
Exact		**5.3**
Pure cells	Air	5.2
Δu-PR	Vacuum	4.8
ACM-1	Vacuum	5.1
ACM-2	Vacuum	5.1
ACM-1 (oblique mesh)	Vacuum	5.1
Δu-PR	Air	4.7
ACM-1	Air	5.1

Table 7. Theoretical and calculated values of longitudinal elastic wave velocity in a plate.

6.4 Discussion of results and conclusions

The calculated data presented here and not included in this work demonstrate that all the methods under consideration have good convergence (the order of convergence is ~1) to the exact solution with mesh refinement as applied to all 1D problems with interfaces. When comparing the methods, one should note that the order of convergence of calculations with closure methods is mostly governed by the order of convergence of the basic difference scheme. As for the error of the closure methods themselves, it is basically controlled by the value of the factor A in formula (106). The reader himself can choose the method he likes. However, two circumstances need to be mentioned, which are important when choosing the method. First, the methods differ in the amount of calculations. Second, the methods differ in the complications in program implementation associated with limitations in their applicability.

As for the 2D problem, the anisotropic methods have no alternative. They possess the same accuracy as the basic methods on the 1D problems, because they rest upon the same closure models, and are more accurate as applied to the 2D problem. Of the two anisotropic methods, it is worth giving preference to ACM-1, because it is easier to implement.

Acknowledgments

The author sincerely thanks Goncharov E, Kolobyanin V, and Toporova O for the joint work on the closure models, Krayukhin A for problem 3 he proposed, and Kamm J and Shashkov M for consultations in the course of the work on the closure models and the setup of the 1D test problems.

Author details

Yury Yanilkin

Address all correspondence to: yan@md08.vniief.ru

VNIIEF, Sarov, Russia

References

[1] Bakhrakh SM, Glagoleva YuP, Samigulin MS, Frolov VD, Yanenko NN, Yanilkin YuV. Gas dynamic flow simulations by the concentration method. DAS USSR. 1981; **257**, N3: 566–569 (in Russian).

[2] Despres B, Lagoutiere F. Numerical solution of two-component compressible fluid model with interfaces. Progress in Computational Fluid Dynamics. 2007; **7**: 295–310.

[3] Shashkov M. Closure models for multimaterial cells in arbitrary Lagrangian-Eulerian hydrocodes. International Journal of Numerical Methods in Fluids. 2007; **56**: 1497–1504.

[4] Noh WF, Woodward P. SLIC (simple line interface calculation). In: van der Vooren, A. I. and Zandbergen, P. J. editors. Numerical Methods for Fluid Dynamics. Berlin: Springer-Verlag; 1976. pp. 330–340.

[5] Yanilkin YV. Numerical modeling of two-dimensional multi-material flows including some sub-scale processes. Physical Mesomechanics. 1999; **2, 5**: 27–48 (in Russian).

[6] Hirt CW, Nicols BD. Volume of fluid (VOF) method for the dynamics of free boundaries. Journal of Computational Physics. 1981; **39**: 201–225.

[7] Youngs DL. Time dependent multi-material flow with large fluid distortion. In: Morton, K.W. and Baines, M. J., editors. Numerical Methods for Fluid Dynamics. New York, NY, USA: Academic Press; 1982. pp. 273–285.

[8] Pillord JE, Pucke EG. Second order accurate volume-of-fluid algorithms for tracing material interfaces. Journal of Computational Physics. 2004; **199**: 465–502.

[9] Dyadechko V, Shashkov M. Reconstruction of multi-material interfaces from moment data. Journal of Computational Physics. 2008; **227**: 5361–5384.

[10] Kucharic M, Garimella RV, Schofield SP, Shashkov M. A comparative study of interface reconstruction methods for multi-material ALE simulations. Journal of Computational Physics. 2010; **229**: 2432–2452.

[11] Hill R, Shashkov M. The symmetric moment-of-fluid interface reconstruction algorithm. Journal of Computational Physics. 2013; **249**: 180–184.

[12] Harlow FH. The particle-in-cell computing method for fluid dynamics. In Methods in Computational Physics Advances in Research and Applications, Volume 3, Fundamental Methods in Hydrodynamics. Academic Press, New York and London, 1964.

[13] Tipton R. CALE mixed zone pressure relaxation. Tech. rep., Lawrence Livermore National Laboratory, Private Communication (1989).

[14] Delov VI, Sadchikov VV. Comparison of several models for computation of thermodynamical parameters for heterogeneous Lagrangian cells. VANT (Mathematical Modeling of Physical Processes). 2005; **1**: 57–70 (in Russian).

[15] Kamm JR, Shashkov MJ. Pressure relaxation closure model for one-dimensional, two-material Lagrangian hydrodynamics based on the Riemann problem. Communications in Computational Physics. 2010; **7**: 927–976. doi:10.4208/cicp.2009.09.032.

[16] Yanilkin Y, Goncharov E, Kolobyanin V, Sadchikov V, Kamm J, Shashkov M, Rider W. Multi-material pressure relaxation methods for Lagrangian hydrodynamics. Computers & Fluids; 2013; **83**: 137–143.

[17] Bakhrakh SM, Spiridonov VF, Shanin AA. A method for 2D axially symmetric heterogeneous flow simulations in Lagrange-Eulerian variables. DAS USSR. 1984; **257**: N4 (in Russian).

[18] Bondarenko, Y, Yanilkin Y. Computation of the thermodynamic parameters in the mixed cells in gas dynamics. Mathematical Modeling. 2002; **14**: 63–81.

[19] Goncharov, EA, Kolobyanin, V, Yanilkin YV. A closure method for Lagrangian hydrodynamics in mixed cells based on equal velocities of material. VANT (Mathematical Modeling of Physical Processes). 2006; **4**: 100–105 (in Russian).

[20] Goncharov EA, Yanilkin YV. A new method for computations of thermodynamic states of materials in mixed cells. VANT (Mathematical Modeling of Physical Processes). 2004; **3**: 16–30 (in Russian).

[21] Barlow A. A new Lagrangian scheme for multimaterial cells. In Proceedings of European Congress on Computational Methods in Applied Sciences and Engineering. ECCOMAS Computational Fluid Dynamics Conference; 2001. pp. 235–294.

[22] Kamm J, Shashkov M, Fung J, Harrison A, Canfield T. A comparative study of various pressure relaxation closure models for one-dimensional two-material Lagrangian hydrodynamics. International Journal of Numerical Methods in Fluids. 2010; **65** (11–12): 1311–1324.

[23] Kamm JR, Shashkov MM., Rider WJ. A new pressure relaxation closure model for one-dimensional two-material Lagrangian hydrodynamics. European Physics Journal Web of Conferences. 2011; **10**. doi:10.1051/epjconf/201010000038.

[24] Hill RN, Barlow A, Shashkov M. Interface-aware sub-scale dynamics closure model. Tech. Rep. LA-UR-12-21959. International Conference on Numerical Methods in Multiphase Flows; 2012-06-12 to 2012-06-14; State College, Pennsylvania, United States and 3rd International EULAG Workshop on Eulerian/Lagrangian methods for fluids. 25–28 June, 2012, Loughborough, UK.

[25] Barlow A, Hill R, Shashkov M. Constrained optimization framework for interface-aware sub-scale dynamics closure model for multimaterial cells in Lagrangian and arbitrary Lagrangian-Eulerian hydrodynamics. Journal of Computational Physics. 2014; **276**: 92–135. doi:10.1016/j.jcp.2014.07.031.

[26] Miller D, Zimmerman G. An algorithm for time evolving volume fractions in mixed zones in Lagrangian hydrodynamics calculations. Tech. rep., Lawrence Livermore National Laboratory, 2006; UCRL-PRES-223908.

[27] Baer M, Nunziato J. A two-phase mixture theory for the deflagration-to-detonation transition (ddt) in reactive granular materials. International Journal of Multiphase Flow. 1986; **12**: 861–889.

[28] Murrone A, Guillard H. A five equation reduced model for compressible two phase flow problems. Journal of Computational Physics. 2005; **202**: 664–698.

[29] Francois M, Shashkov M, Dendy E, Lowrie R. Mixture models for multimaterial Eulerian and Lagrangian hydrocodes. Tech. rep., Presentation at 8th International Conference on New Models and Hydrocodes for Shock Wave Processes, Paris, France; Available as Los Alamos National Laboratory Report LAUR-10-03391, 2010.

[30] Barlow A. A new Lagrangian scheme for multimaterial cells. In Proceedings of European Congress on Computational Methods in Applied Sciences and Engineering, ECCOMAS Computational Fluid Dynamics Conference; Swansea, Wales, U.K. 2001. pp. 235–294.

[31] Grove JW. Pressure-velocity equilibrium hydrodynamic models. Acta Mathematica Scientia. 2010; **30B(2)**: 563–594.

[32] Sun M. A thermodynamic and dynamic subgrid closure model for two-material cells. International Journal of Numerical Methods in Fluids. 2013; **73(2)**: 130–151.

[33] Rider W, Love E, Wong M, Strack O, Petney S, Labreche D. Adaptive methods for multi-material ALE hydrodynamics. International Journal of Numerical Methods in Fluids. 2011; **65(11–12)**: 1325–1337.

[34] Yanilkin YV, Goncharov EA, Toporova OO, Kolobyanin VY. Closure methods for the lagrangian gas dynamics and elasto-plasticity equations in mixed cells. Presentation at International Conference on New Models and Hydrocodes for Shock Wave Processes, PETER 2016, St Malo, France, 30 May–3 June 2016.

[35] Yanilkin YV, Belyaev SP, et al. EGAK and TREK: Eulerian codes for multidimensional multi-material flow simulations. RFNC-VNIIEF Transactions. Research Publication. Sarov: RFNC-VNIIEF. 2008; **12**: 54–65 (in Russian).

[36] Goncharov EA, Kolobyanin VY, Yanilkin YV. On the determination of artificial viscosity for materials in mixed cells. VANT (Mathematical Modeling of Physical Processes). 2010; **2**: 15–29 (in Russian).

[37] Zharova GV, Yanilkin YuV. EGAK code. Pressure equilibration algorithm for mixed cells. VANT (Mathematical Modeling of Physical Processes). 1993; **3**, 77–81 (in Russian).

[38] Benson DJ. Computational methods in Lagrangian and Eulerian hydrocodes. Computer Methods in Applied Mechanics and Engineering. 1992; **99(23)**: 235–235. doi:10.1016/0045-7825(92)90042-I.

[39] Weseloh WN, Glancy SP, Painter JW. PAGOSA Physics Manual. Tech. Rep. LA-14225-M; Los Alamos National Laboratory; 2010.

[40] Yanilkin YV. Study and implementation of multi-material pressure relaxation methods for Lagrangian hydrodynamics. Tech. Rep. LA-UR-10-06664; Los Alamos National Laboratory; 2010.

[41] Yanilkin YV, Statsenko VP, Kozlov VI. Mathematical modeling of turbulent mixing in compressible matter. A course of lectures. FSUE RFNC-VNIIEF Publishing Office, Sarov; 2009 (in Russian).

[42] Krayukhin AA, Svidinsky VA, Stadnik AL, Yanilkin YV. Non-steady-state problems for testing elastoplastic techniques. VANT (Mathematical Modeling of Physical Processes). 2016; **2**: 17–30 (in Russian).

[43] Saurel R, R Abgrall. A Multiphase Godunov Method for Compressible Multifluid and Multiphase Flows. Journal of Computational Physics. 1999; **150**: 425–467.

[44] Plohr B. Shockless acceleration of thin plates modeled by a tracked random choice method. AIAA Journal. 1988; **26**: 470–478.

Chapter 5

Mechanics of Electric Rope Shovel Performance and Reliability in Formation Excavation

Muhammad Azeem Raza and Samuel Frimpong

Additional information is available at the end of the chapter

http://dx.doi.org/10.5772/65333

Abstract

Large-capacity rope shovels are used as primary production equipment in many surface mining operations. Current rope shovels have payload capacities in excess of 100 tons per pass. The dynamic payload and formation resistive forces result in severe stress loading of the shovel front-end assembly. Material flaws, high stresses, and harsh excavation conditions can initiate cracks in the dipper-teeth assembly. These cracks, under high stresses, can propagate to critical lengths resulting in fatigue failure of front-end assembly. Dipper-related problems can significantly reduce shovel availability. There is no fundamental research for understanding dipper fatigue failure resulting from high stress intensity, crack initiation, and propagation, the subject matter of this study. The Newton-Euler algorithm is used to build kinematics and dynamic models of the cable shovel front-end assembly. The models incorporate the dynamic resistive forces on the dipper-teeth assembly. Numerical simulations are used to generate the dynamic payload force and its dynamic left. Virtual simulation, based on the P&H 4100XPC shovel prototype in ANSYS (R15), is run to generate stress loading of the dipper-teeth assembly and equivalent (von Mises) stresses. Stress intensity factors are computed for various crack lengths in the dipper-teeth assembly, and the crack-propagation lives are computed for these cracks. The results show that a 75-mm crack can propagate to the critical length in 16 days. This research study provides a pioneering effort toward understanding shovel dipper fatigue failure due to high stress intensity, crack initiation, and propagation for understanding shovel reliability and availability for production efficiency and bulk production economics.

Keywords: formation excavation, machine kinematics and dynamics, virtual simulation, stress profile intensity

1. Introduction

Cable shovels are used as primary excavation equipment in large-scale surface mining operations. The overall efficiency of shovel-truck surface mining operations is largely dependent on shovel efficiency. Dipper payloads of the shovels have seen an increasing trend over the years, and current shovels have payloads in excess of 100 tons per scoop [1, 2]. The payloads, combined with dipper weight, rigging, and variable material diggability, result in varying mechanical energy inputs and stress loading of the shovel's front-end assembly across the working bench. Furthermore, the repeated shovel loading and unloading cycles induce fatigue stresses in shovel components. The induced stresses over time may exceed the yield strength of steel/material of the shovel leading to fatigue failure, teeth losses, and boom and handle cracks. High stresses and fatigue failure in shovel front-end assembly cause unplanned downtimes resulting in reduced efficiency and increased production costs. Dipper-related problem can be a significant contributor to the shovel downtime [3]. The current practice for the shovel front-end assembly repair is based on experience and history rather than science.

Electric rope shovel consists of the lower, upper, and the front-end assembly as illustrated in **Figure 1**. The lower assembly consists of the propel drive and crawler systems and provides a solid and stable base for the excavator. This helps excavator propel, reposition, and relocation during its operation.

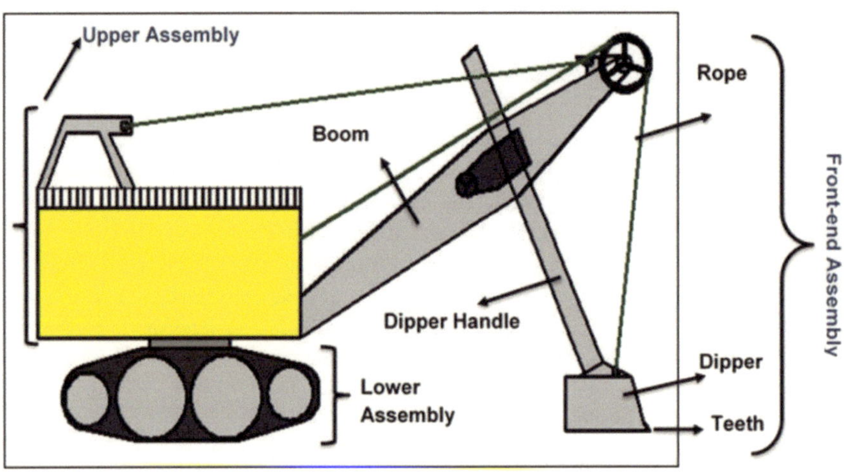

Figure 1. Nomenclature of a cable shovel.

The shovel's upper assembly is a roller and left-pin system mounted on the lower mechanism. The upper assembly consists of multiple decks with housing for the hoist and swing machinery and electronic control cabinet on the lower deck and the operator's cab on the upper deck. Additionally, the upper assembly provides a platform for boom attachment and the counter weight for the dipper. The front end consists of the boom, crowd machinery, dipper handle,

http://dx.doi.org/10.5772/65333

dipper, and ropes. During the normal duty cycle, the shovel stays at one position, and only the front end engages with the formation.

2. Shovel resistance forces and modeling

The dipper excavation processes can be categorized into penetration, cutting, and scooping processes [4, 5]. Penetration is the insertion of a tool into a medium, and cutting is the lateral movement of a tool, executed at a constant depth. The dipper teeth penetrate the formation, and the lip cuts the material. Excavation models are based on the formation resistive forces acting on the cutting tool. The resistive forces combine the cutting forces at the dipper teeth and lip and the excavation forces due to material movement along, ahead, and inside the dipper. Both the experimental and analytical models are based on these resistive forces. The model proposed by Hemami [6] is by far the most comprehensive model and consists of six component forces (f_1 to f_6), which must be overcome during excavation, as in **Figure 2**. All these forces, except f_6, are dynamic forces. The six forces acting on the dipper, from the initial to the end point on trajectory, consist of the following:

f_1: The force required to overcome the payload weight in and above the dipper

f_2: The resultant resistive force due to material movement toward the dipper

f_3: The friction force between the bucket walls and the excavated material as it slides into the dipper

f_4: The resistance to cutting and/or penetrating that acts at the dipper tip and side walls

f_5: The inertia force of the material inside and above the dipper

f_6: The force required to move the empty dipper (modeled as part of f_1)

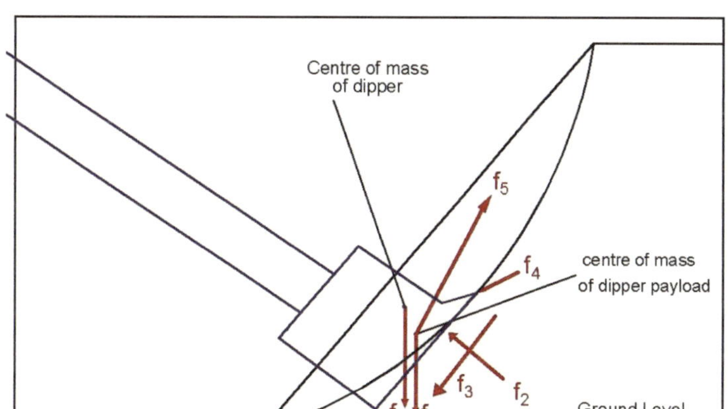

Figure 2. Forces on a dipper during excavation [7].

The forces, f_1 and f_5, are the dynamic forces [7], where f_1 changes both in magnitude and the point of application, and f_5 depends on the bucket acceleration. The force (f_6) was originally defined as a part of f_1 and f_5. The dipper payload force (f_1) is the dominant force for the large-capacity dippers [6, 8, 9]. Awuah-Offei et al. [8] proposed a model based on the Balovnev [10] excavation model using the six forces. The force (f_2) can be set to zero [6]. Forces, f_3 and f_4, are the cutting forces and can be combined as a single force and estimated using the empirical model [11] given by Eq. (1).

This empirical model is a result of extensive experimentation on frozen soils [11]. z is the coefficient that accounts for the blade impact on cutting force, which depends on w and d. **Table 1** is used to estimate z for d (between 25 cm and 50 cm). z increases as d decreases, and it also depends on the ratio T_s/T_w (T_s is the spacing between the teeth, and T_w is the tooth width). **Table 2** lists the multiplying factors for z based on T_s/T_w. Force f_5 can also be set to zero if the dipper moves with a uniform velocity through the muck pile. Force f_6 can be modeled as part of f_1:

Length of horizontal surface (w, m)	0.25–0.50	0.50–0.75	0.75–1.00	1.00–1.25
Coefficient z	0.55–0.75	0.63–0.78	0.69–0.8	0.71–0.82

Table 1. Dependence of "z" on "d," and "w".

Ratio T_s/T_w	$T_s = T_w$	$T_s = 2T_w–3T_w$	$T_s = 4T_w$	$T_s = 5T_w$
z	1.2	1	1.1	1.25

Table 2. Dependence of z on T_s/T_w [11].

$$P=10C_o\ d^{1.35}\ (1+2.6w)(1+0.0075\beta')z \tag{1}$$

3. Kinematic model of the cable shovel front end

A kinematic model of the shovel is required to completely describe the motions (accelerations, velocities, and displacements). The kinematic model further provides a basis for the dynamic model, which can be used to calculate the torques and forces on individual components. The complete shovel digging process involves propel, crowd, and swing motions. However, during the normal duty cycle, the shovel positions itself against the working face without propel. In this situation, only the front-end assembly moves. Further, the maximum forces are involved during the excavation phase. Therefore, a dynamic model of the front-end assembly alone can suffice to describe the normal duty cycle of cable shovel. **Figure 3** shows the shovel front-end assembly, whose mechanism is modeled as a three-link system (saddle, crowd arm, and dipper) with three links and three joints. The saddle is a fixed length link and is free to rotate in the vertical plane. The rotation of the saddle block controls the vertical position of the dipper. The crowd arm is connected to the saddle block through a prismatic joint, and its length varies during the crowding action of the digging operation.

The length of the crowd arm controls the horizontal position of the dipper. The crowd arm and the saddle have the same rotation cycle, while the dipper is oriented at a fixed angle, β, to the crowd arm. The dipper is also a fixed length link. The rotation of the saddle block and the length of the crowd arm together control the position of the dipper in the vertical plane and its trajectory. The structural kinematic parameters of the shovel using the Denavit-Hartenberg (D-H) notation [12] are represented in **Figure 3** and **Table 3**. Here, four values are assigned to each link following the D-H notation.

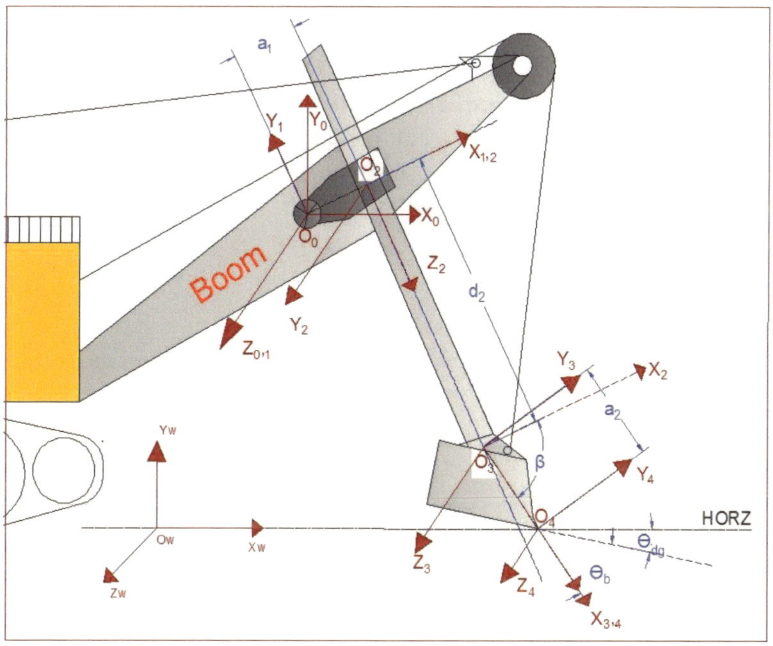

Figure 3. Structural kinematics using D-H procedure.

Link i	Joint description	α_i	a_i	d_i	θ_i
1	Saddle-boom joint	0	0	0	θ_1
2	Saddle-dipper handle joint	90	a_1	0	0
3	Dipper handle-dipper joint	−90	0	d_2	0

Table 3. Structural kinematic parameters.

The two values (ai, di) are for the links and represent the constant and variable lengths of the links, while the other two (αi, θi) are for the connection between links (i.e., joints), and, thus, represent the rotation of the coordinate frame and rotation of the joint, respectively. For a revolute joint, ai, αi, and di are fixed and θi is a variable. On the other hand, for a prismatic joint (or translational motion), ai, αi, and θi are fixed and di is a variable. The crowd-arm movement is via a prismatic joint. A kinematic scheme relates the movements of the links and translates the motions and rotations in the reference coordinate frame. The D-H scheme is used to relate the movements and rotation of the links. The movements and rotations of individual links are measured in the coordinate frames assigned at every joint location using the D-H scheme [12]. The lower part of the shovel is stationary and fixed for this analysis.

The $X_oY_oZ_o$ frame, the reference coordinate frame, is selected with Z_o along the rotating axis of the saddle block. The coordinate frame $X_1Y_1Z_1$ coincides with the $X_oY_oZ_o$ frame and measures the rotation of the dipper handle via the saddle block. Next, the coordinate frame $X_2Y_2Z_2$ is set at the intersection of the saddle block and the dipper handle, with the Z_2 axis along the translation movement of the dipper handle (joint 2 being a prismatic joint). The movement of the dipper handle is measured along this Z_2 axis. The coordinate frame $X_3Y_3Z_3$ is set at the end point of the dipper handle with Z_3 normal to Z_2. This frame is at a fixed angle from coordinate frame 2. And finally, the frame $X_4Y_4Z_4$ is set at the tip of the dipper with Z_4-axis parallel to Z_3. The material resistive forces acting on the shovel are defined in this frame. The coordinate frame assignments are also shown in **Figure 3**.

Forward kinematics of cable shovel front-end assembly: The forward kinematic model defines the positions and motions of the dipper with known dipper-handle rotation and extension. External dynamic forces act on the shovel dipper during excavation. A transformation scheme is used to translate point coordinates in one coordinate frame to the first coordinate frame. The homogenous transformation matrix for transferring coordinates from *i–1* coordinate frame to *i* frame, in its general form for revolute and prismatic joints is given in Eqs. (2) and (3), respectively [13]. These equations can be derived considering two links (*i–1* and *i*) connected through revolute or prismatic joints, respectively. These transformation equations are a combination of rotation and translation matrices:

$$T_i^{i-1} = \begin{bmatrix} \cos\theta_i & -\cos\alpha_i\alpha_i\sin\theta_i & \sin\alpha_i\sin\theta_i & a_i\cos\theta_i \\ \sin\theta_i & \cos\alpha_i\cos\theta_i & -\sin\alpha_i\cos\theta_i & a_i\sin\theta_i \\ 0 & \sin\alpha_i & \cos\alpha_i & d_i \\ 0 & 0 & 0 & 1 \end{bmatrix} \quad (2)$$

$$T_i^{i-1} = \begin{bmatrix} \cos\theta_i & -\cos\alpha_i\alpha_i\sin\theta_i & \sin\alpha_i\sin\theta_i & 0 \\ \sin\theta_i & \cos\alpha_i\cos\theta_i & -\sin\alpha_i\cos\theta_i & 0 \\ 0 & \sin\alpha_i & \cos\alpha_i & d_i \\ 0 & 0 & 0 & 1 \end{bmatrix} \quad (3)$$

The individual transformation matrices T_i^{i-1} are formulated using Eqs. (2) and (3). These matrices relate the geometry of a point in the two adjacent coordinate frames as in **Figure 3** and can further be multiplied together to obtain a transformation matrix between any two coordinate frames. These transformations are required for the shovel front-end kinematic and dynamic models using the Newton-Euler procedure. The Newton-Euler method is an iterative method for computing the velocities, accelerations, joint torques, and forces from crowd arm to dipper in the forward direction and from dipper tip to the saddle block in the reverse direction. Newton-Euler method has an advantage of being iterative, which makes it more suitable for computer simulations.

The propagation of angular and linear velocities from joint to joint is given by Eqs. (4) through (7) [13]. For rotational motion, the angular and linear velocities are defined by Eqs. (4) and (5), respectively. For prismatic joint, the angular and linear velocity relations are given by Eqs. (6) and (7), respectively:

$$^{i+1}\omega_{i+1} = {}^{i+1}_{i}R\,{}^{i}\omega_{i} + \dot{\theta}_{i+1}\ {}^{i+1}\hat{Z}_{i+1} \tag{4}$$

$$^{i+1}v_{i+1} = {}^{i+1}_{i}R\,({}^{i}v_{i} + {}^{i}\omega_{i}\ X\ {}^{i}P_{i+1}) \tag{5}$$

$$^{i+1}\omega_{i+1} = {}^{i+1}_{i}R\,{}^{i}\omega_{i} \tag{6}$$

$$^{i+1}v_{i+1} = {}^{i+1}_{i}R\,({}^{i}v_{i} + {}^{i}\omega_{i}\ X\ {}^{i}P_{i+1}) + \dot{d}_{i}\hat{Z}_{i} \tag{7}$$

The required rotation matrices are derived from the transformation matrices in Eqs. (2) and (3). The 3x3 matrix, within a transformation matrix T_i^{i-1}, represents the corresponding rotation matrix R_i^{i-1}. The forward kinematic starts from the first link (saddle block) and moves outward toward the last link (dipper). The objective is to determine the propagation of the joint rotation and velocities from the joint 1 to the dipper tip. The model uses the same start point equations and basic simplifying assumption from Frimpong et al. [14], and as a result, the kinematic equations are very similar as well. However, the resulting dynamic model is different due to the improved resistive forces in this model. The reference frame {0} is fixed with the lower frame through the boom. The lower structure of the shovel is fixed, so its linear and angular velocities and accelerations remain zero at all times during the excavation as shown in Eqs. (8) and (10). These values change only during the propel motion of shovel which is not considered in this research. The joint velocity can be determined by taking the derivative of rotation of joint 1 as shown in Eqs. (9) and (11), respectively. Similarly, the linear velocity of the stationary lower structure of the shovel is zero:

$$^{0}\omega_{0} = \begin{bmatrix} 0 \\ 0 \\ 0 \end{bmatrix} \tag{8}$$

$$\frac{d\,{}^{0}\omega_{0}}{dt} = {}^{0}\dot{\omega}_{0} = \begin{bmatrix} 0 \\ 0 \\ 0 \end{bmatrix} \tag{9}$$

$${}^{0}v_{0} = \begin{bmatrix} 0 \\ 0 \\ 0 \end{bmatrix} \tag{10}$$

$$\frac{d\,{}^{0}v_{0}}{dt} = {}^{0}\dot{v}_{0} = \begin{bmatrix} 0 \\ 0 \\ 0 \end{bmatrix} \tag{11}$$

Eq. (12) is obtained from Eq. (4) for joint 1 ($i = 0$), a revolute joint. It is evident from this equation that the angular velocity of the first link is only around Z-axis and is equivalent to the rate of change of angular rotation around joint 1. The linear velocity propagation to joint 1 can be computed using Eq. (5) as Eq. (13). The first link experiences only the rotational motion. Therefore, the linear velocity of joint 1 is zero:

$${}^{1}\omega_{1} = {}^{1}_{0}R\,{}^{0}\omega_{0} + \dot{\theta}_{1}\,{}^{1}\hat{Z}_{1} = \omega_{1} = {}^{1}_{0}R\,{}^{0}\omega_{0} + \dot{\theta}_{1}\,{}^{1}\hat{Z}_{1} = \begin{bmatrix} 0 \\ 0 \\ \dot{\theta}_{1} \end{bmatrix} \tag{12}$$

$${}^{1}\mathrm{v}_{1} = {}^{1}_{0}\mathrm{R}\,({}^{0}\mathrm{v}_{0} + {}^{0}\omega_{0}\ \mathrm{X}\ {}^{0}\mathrm{P}_{1}) = \begin{bmatrix} 0 \\ 0 \\ 0 \end{bmatrix} \tag{13}$$

For the prismatic joint 2 ($i = 1$), Eq. (6) computes the angular velocity of the link 2 (the crowd arm). The propagation of angular velocity to joint 2, as given in Eq. (14), shows that the angular velocity of joint 2 is dependent upon the rate of change of angular rotation of joint 1, and there is only an axis shift involved (from Z-axis to Y-axis) during the propagation:

$${}^{2}\omega_{2} = {}^{2}_{1}R\,{}^{1}\omega_{1} + \dot{\theta}_{2}\,{}^{2}\hat{Z}_{2} = \begin{bmatrix} 0 \\ \dot{\theta}_{1} \\ 0 \end{bmatrix} \tag{14}$$

$$\therefore \theta_{2},\ \dot{\theta}_{2}, \text{and}\ \ddot{\theta}_{2}\ \textit{are zero}$$

$$ {}^{2}v_{2} = {}^{2}_{1}R({}^{1}v_{1} + {}^{1}\omega_{1} \times {}^{1}P_{2}) = \begin{bmatrix} 0 \\ 0 \\ -a_{1}\dot{\theta}_{1} \end{bmatrix} \tag{15} $$

The linear velocity propagation to joint 2 is calculated as Eq. (15). Similarly, the angular and translational velocities are calculated for joint 3 as Eqs. (16) and (17). Again, the angular velocity of joint 3 is equivalent to the rate of change of angular rotation of joint 1. There is only one rotation of the joint involved for the front end during the digging cycle. Thus, the angular velocity of joint 4 is also the same as the angular velocity of joint 1. Alternately, it can be stated that the whole front-end assembly gets the same rotation as the joint 1 during the digging cycle, and the angular velocity only involves the axis shift. Eqs. (8) through (18) define the forward kinematics of the shovel front end. The angular and linear velocities of the shovel front-end components are defined using these equations with known initial rotation and crowd-arm extension:

$$ {}^{3}\omega_{3} = \begin{bmatrix} 0 \\ 0 \\ \dot{\theta}_{1} \end{bmatrix} \tag{16} $$

$$ {}^{3}v_{3} = {}^{3}_{2}R({}^{2}v_{2} + {}^{2}\omega_{2} \times {}^{2}P_{3}) = \begin{bmatrix} d_{2}\dot{\theta}_{1} \\ a_{1}\dot{\theta}_{1} \\ 0 \end{bmatrix} \tag{17} $$

$$ {}^{4}v_{4} = {}^{4}_{3}R({}^{3}v_{3} + {}^{3}\omega_{3} \times {}^{3}P_{4}) = \begin{bmatrix} d_{2}\dot{\theta}_{1}c_{\beta} + (a_{1} + a_{2})\dot{\theta}_{1}s_{\beta} \\ -d_{2}\dot{\theta}_{1}s_{\beta} + (a_{1} + a_{2})\dot{\theta}_{1}c_{\beta} \\ 0 \end{bmatrix} \tag{18} $$

$$ T_{4}^{1} = \begin{bmatrix} c_{\beta} & -s_{\beta} & 0 & a_{1} + a_{2}c_{\beta} \\ s_{\beta} & c_{\beta} & 0 & a_{2}s_{\beta} - d_{2} \\ 0 & 0 & 1 & 0 \\ 0 & 0 & 0 & 1 \end{bmatrix} \tag{19} $$

Inverse kinematics of cable shovel front-end assembly: The inverse shovel kinematics determine the set of joint angles and the length for the crowd arm when the desired position and orientation of the shovel dipper are known in the reference coordinate frame 0. This inverse kinematic is useful when the dipper traverses a known trajectory to determine the joint rotation and crowd-arm extension required to achieve this trajectory. An approach, similar to the one used by Wu [15] for the reverse kinematic model of cable shovel, is used to determine the crowd-arm extension and rotation with known trajectory points. The inverse kinematic model can be achieved by coordinate transformations to obtain the dipper coordinate in coordinate frame

4, relative to coordinate frame 1. The modifications of the transformation matrix equations result in Eqs. (19), (20), and (21):

$$T_4^1 = \left[T_1^0\right]^{-1} T_4^0 \tag{20}$$

$$T_4^0 = \begin{bmatrix} r_{11} & r_{12} & r_{13} & p_x \\ r_{21} & r_{22} & r_{23} & p_y \\ r_{31} & r_{32} & r_{33} & p_z \\ 0 & 0 & 0 & 1 \end{bmatrix} \tag{21}$$

The individual matrix elements are given as follows:

$$\begin{aligned}
&r_{11} : c_1 c_\beta - s_1 s_\beta; \\
&r_{12} : -c_1 s_\beta - s_1 c_\beta; \\
&r_{13} : 0; \\
&r_{21} : s_1 c_\beta + c_1 s_\beta; \\
&r_{22} : -s_1 s_\beta + c_1 c_\beta; \\
&r_{23} : 0; \\
&r_{31} : 0;\ r_{32} : 0; \\
&r_{33} : 0; \\
&P_x : a_2 c_1 c_\beta - a_2 s_1 s_\beta + a_1 c_1 + d_2 s_1; \\
&P_y : a_2 s_1 c_\beta + a_2 c_1 s_\beta + a_1 s_1 - d_2 c_1; \\
&P_z : 0
\end{aligned}$$

Here, (p_x, p_y, and p_z) are the coordinates of the dipper tip in the reference coordinate frame 0. Eq. (22) can be derived from Eq. (2) and Eq. (23) from Eq. (20). Comparing the individual matrix elements on both sides of Eq. (23) and using simple arithmetic and trigonometric operations, the crowd-arm extension and rotation can be computed using Eqs. (25) and (26), respectively. The inverse kinematic model can be used to compute the positions and velocities of individual links and joints of the front-end assembly for a known trajectory:

$$\left[\mathrm{T}_1^0\right]^{-1} = \begin{bmatrix} \mathrm{c}_1 & \mathrm{s}_1 & 0 & 0 \\ -\mathrm{s}_1 & \mathrm{c}_1 & 0 & 0 \\ 0 & 0 & 1 & 0 \\ 0 & 0 & 0 & 1 \end{bmatrix} \tag{22}$$

$$\begin{bmatrix} - & - & - & c_1 p_x + s_1 p_y \\ - & - & - & -s_1 p_x + c_1 p_y \\ - & - & - & p_z \\ - & - & - & 1 \end{bmatrix} = \begin{bmatrix} c_\beta & -s_\beta & 0 & a_1 + a_2 c_\beta \\ s_\beta & c_\beta & 0 & a_2 s_\beta - d_2 \\ 0 & 0 & 1 & 0 \\ 0 & 0 & 0 & 1 \end{bmatrix} \tag{23}$$

$$d_2 = a_2 s_\beta + \sqrt{p_x^2 + p_y^2 - a_1^2 - a_2^2 - 2a_1 a_2 c_\beta + \left(a_2 s_\beta\right)^2} \tag{24}$$

$$\theta_1 = A\tan 2\left(a_2 s_\beta - d_2 , \pm\sqrt{p_x^2 + p_y^2 - \left(a_2 s_\beta - d_2\right)^2} \right) - A\tan 2(p_y, p_x) \tag{25}$$

4. Dynamic model of the cable shovel front-end assembly

The dynamic model defines forces and torques acting on the shovel links and joints from the kinematics parameters, such as accelerations. The forces require the computation of angular and linear accelerations, which can be obtained by time integration of the angular and linear velocities computed in the kinematic model. In its general form, the dynamic model can be defined as in Eq. (26) from Frimpong et al. [16]:

$$\begin{gathered} D(\Theta)\ddot{\Theta} + C(\Theta,\dot{\Theta})\dot{\Theta} + G(\Theta) = F - F_{load}(F_t, F_n) \\ D(\Theta) = \text{mass matrix} \\ C(\Theta,\dot{\Theta}) = \text{centrifugal and Coriollis terms} \\ G(\Theta) = \text{gravity terms} \end{gathered} \tag{26}$$

This dynamic model for a shovel is built using the Newton-Euler method and the position, velocity, and acceleration relationships computed from the kinematic model. The Newton-Euler dynamic algorithm for computing the crowd force and the hoist torque comprises of the following steps:

1. Compute the angular acceleration ($\dot{\omega}_i$) of every link in the forward direction, starting from the saddle and moving outward toward the last link (the dipper).

2. Compute the acceleration ($\dot{v}_i$) of every link in the system in the forward direction.

3. Compute the acceleration ($\dot{v}_{ic}$) at the left of mass (centroid) of every link in the system in the forward direction.

4. Determine the force (F_i) acting on every link at the centroid of the link using ($\dot{v}_{ic}$) and mass of the link.

5. Compute the joint torque (N_i) for every link.

The force and torque are computed at the centroid of each link. Therefore, the velocity and the acceleration of the centroid are computed for every link.

5. Numerical modeling and simulation of the dipper-formation interaction

The dynamic model is a system of ordinary differential equations (ODEs), which results from an iterative process and includes a number of ODE subprocesses. The ODEs are numerically solved in MATLAB using the embedded Runge-Kutta algorithm. The simulation model consists of MATLAB programs (.m files) and SIMULINK design-based models and sub-models. The simulation model consists of the main model and sub-models. These sub-models define the dipper's trajectory, the crowd-arm extension and rotation, and the resistive forces (cutting forces, material, and dipper's weight) on the dipper. The following sub-models and main model are created:

1. *Test bench geometry and trajectory*: **Figure 4** shows the test bench geometry created for the digging process simulation. The excavated material characteristics can be selected for various digging conditions. A simulation step size is selected to make the dipper move with a constant linear velocity following field experimental results [17]. The failure surface is modeled as a quadratic function given by Eq. (27), and bench face is modeled as a straight-line function L(x). During the simulation process, the coordinates of the dipper tip (O_4(x,y)) and the dipper depth into the working bench (d) are continuously computed at every time step using Eqs. (28) and (29), respectively:

$$y = 0.9927\text{x}^2 - 22.557\text{x} + 117.68 \tag{27}$$

$$O_4x, y = \left[-0.4837t^2 + 2.4351t + 12.053, \qquad 0.9927 * x^2 - 22.557 * x + 117.68\right] \tag{28}$$

$$d = O_4(x) - L(x) \tag{29}$$

http://dx.doi.org/10.5772/65333

2. *Crowd-arm extension and rotation angle*: This sub-model calculates the extension and rotation of the crowd arm at every simulation step using Eqs. (24) and (25).

3. *Payload and force f_1*: The payload forms the basis for the dynamic payload force (f_1). **Figure 4** defines the dipper trajectory in such a way that the dipper is filled as it leaves the bank without any material spillage. At any instant, therefore, the payload is equivalent to the area between the trajectory curve and bench face.

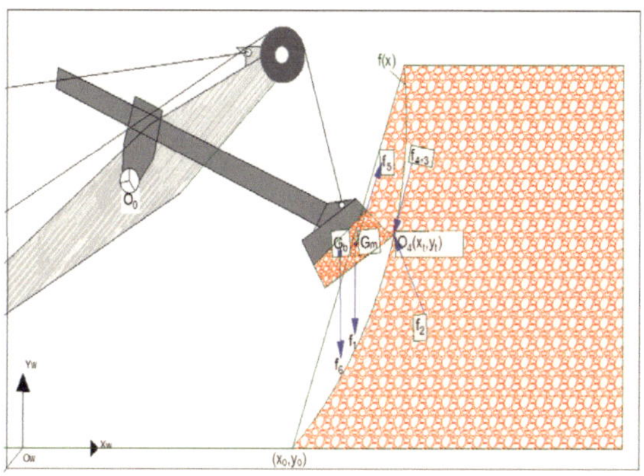

Figure 4. Representative bench geometry.

At each simulation step, the (x, y) trajectory coordinates and the area excavated (A_c) are numerically computed using Eq. (30) [8] and built-in routines in MATLAB R2012a. This area is used to calculate the force (f_1) due to the payload weight using Eq. (31) [8]. An optimization algorithm [8] is used to define the geometry of the payload based on the material distribution from Hemami [6]. The centroid of each material geometry, a polygon inside the dipper, is computed using a special algorithm [18]. This centroid is a dynamic point, which is used as the point of application for the dynamic force f_1. This force (f_1) is computed continuously at every instant of the excavation process:

$$A_c = \frac{1}{2}(x_t - x_o)^2 \tan\alpha - \int_{x_o}^{x_t} f(x)\,dx \tag{30}$$

$$f_1 = A_c \omega \rho g \tag{31}$$

4. *Material resistive force f_6*: The force, due to the weight of the dipper, is calculated during the digging cycle along the trajectory. The computation of centroids of payload geometry and dipper suggests that these two centroids can be considered concentric. Therefore, both f_1 and f_6 are combined into single force acting at the dynamic left of the payload geometry.

5. *Digging resistive forces f_3 and f_4*: The resistive forces, f_3 and f_4, are combined as a single cutting force (F_r) and calculated using Eq. (1) [11]. The cutting force (F_r) acts along the tangent of the trajectory at the dipper tip. This force is resolved into its rectangular components, one along the dipper base and the other normal to it. These tangent and normal components (F_t and F_n) of the resistive force (F_r) are computed at every trajectory point in this sub-model.

6. *The main model and numerical simulation*: The dynamic model of the dipper-teeth assembly is solved in the main model. The outputs from all the sub-models, along with system constants and time steps are fed into the main model as inputs. The main model then numerically solves the mathematical model and generates the desired outputs. Two of the important results or outputs from this solution are the hoisting force (F_1) and crowd-arm torque (T_1).

During this numerical simulation process, four of the six resistive forces (f_1, f_3, f_4, f_6) are computed as separate subsystems, while the other two resistive forces (f_2 and f_5) are set to zero. The resistive force f_2 is set to zero by selecting an appropriate trajectory of the dipper [6]. The excavation trajectory is selected in such a way that the dipper stays clear off the material and does not compress the material. This assumption is reasonable in the sense that it involves proper bench geometric design and operator skill. An improper bench geometric design would lead to undue stresses on the shovel, which must be avoided during the excavation process. The force f_5 represents the dipper and payload inertia. This force can be set to zero if the dipper moves through the material with a constant velocity and hence with zero acceleration. For this research, it is assumed that the dipper moves through the bench with a constant velocity and hence a zero acceleration. This assumption is consistent with the field observations [17] for hoist rope extension.

6. Virtual shovel prototype simulation

A virtual 3D prototype of the rope shovel is built in AutoCAD-2012 as shown in **Figure 5**. The dimensions of the shovel front-end assembly are chosen to represent the dimensions of the P&H 4100XPC shovel and are measured from a scaled model [19]. The front-end geometry is simplified to avoid unnecessary geometric complications. The model consists of one revolute and two prismatic joints that control the motion of the dipper into the formation. The boom and saddle are modeled as rigid bodies. The boom is considered fixed to the ground. Both joints are constrained; the revolute joint allows rotation only in the z-axis, and the prismatic joints allow motion only in the x-axis. The resistive forces of the formation are applicable as a remote force available at the teeth. The material force is also modeled as a remote force acting on the dipper. The revolute joint is given a fixed rotation at every time step to ensure the completion of the digging cycle in 3 s. The contacts and boundary conditions are shown in **Figure 6**. The dipper body is modeled using brick elements with a minimum of three elements through the thickness of the dipper.

http://dx.doi.org/10.5772/65333

The dipper trajectory is given as an input function to the shovel simulation process in MATLAB/SIMULINK. The dipper traverses the known trajectory, and the reverse kinematic model is used to determine the crowd-arm extension (d_2) and rotation (θ_1) requirements to achieve this trajectory. These two output parameters from the numerical simulation process are used as inputs for the shovel prototype. Together, these two inputs define the dipper trajectory. Similarly, the resistive forces computed during the shovel dynamic simulation are modeled as higher order polynomial in MATLAB and are fed into the system as time functions. The payload also exerts a force on the dipper side walls. This force is modeled using the earth pressure at-rest theory [20] and is considered to be acting uniformly over the side wall.

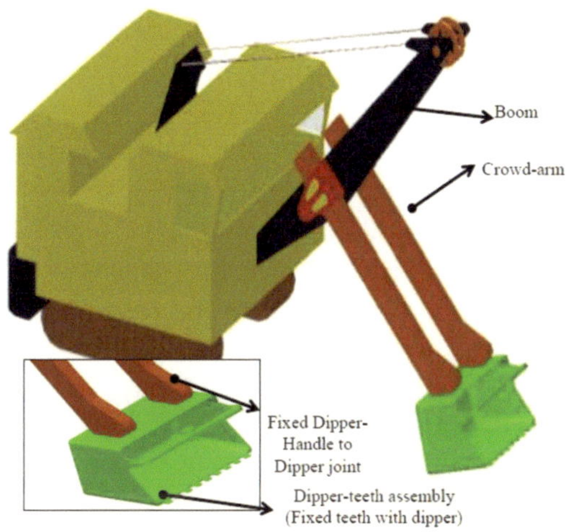

Figure 5. Simplified 3D model of cable shovel and dipper.

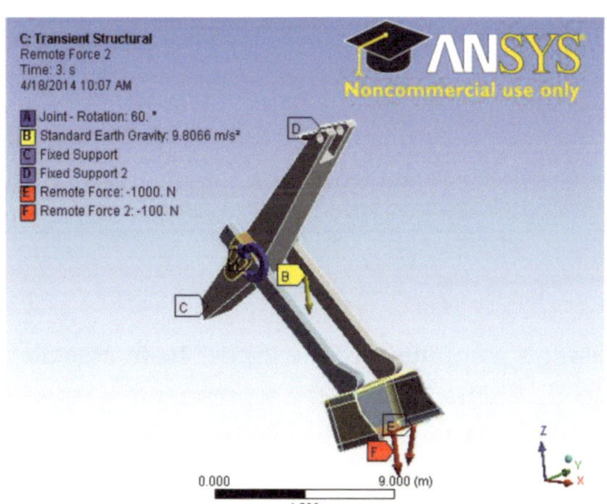

Figure 6. Boundary conditions and external forces on shovel front end.

7. Shovel stress modeling and analysis

The steel used for cable shovel dippers and teeth has high strength properties. The yield strength is the most important strength property, as the shovel experiences high levels of stress loading. Limited data is available for shovel components [21]. One research reported csa-g40.21-350WT steel used for shovel boom [21]. This steel has high yield strength of 320 MPa [21, 22]. It is assumed in this study that the same steel is used for the dipper and the shovel boom, and their properties are given in **Table 4**. The angular rotation and extension of the dipper arm, external digging forces, and dynamic material weight forces from the dynamic simulated model are used as inputs for the virtual prototype. The shovel stress analysis is performed in ANSYS Workbench-R15.0 [30]. First, a rigid-body analysis is performed to ensure that the dipper follows the given trajectory for the given angular rotation, crowd-arm extension, and external forces. Afterward, a transient analysis is performed for the dipper and teeth stress analysis in ANSYS Workbench R15. For this analysis, the dipper-teeth assembly and crowd arm are converted into flexible bodies, allowing ANSYS to compute stresses on the dipper components. All force functions are the same as that used for the rigid-body analysis. The joint functions are defined for the desired trajectory generation. The flexible dipper bodies are meshed appropriately, using sweepable bodies and controlled meshing. The simulations are performed for a 3 s interval. The simulation is run in two steps with multiple sub-steps for better convergence.

Property	Value	Unit
Density	7900	kg/m^3
Young's modulus	2.3E+11	Pa
Poisson's ratio	0.3	
Tensile yield strength	3.2E+08	Pa
Compressive yield strength	3E+08	Pa
Tensile ultimate strength	4.6E+08	Pa

Table 4. Properties of steel for dipper and teeth.

Stress loading (von Mises) is computed for the dipper-teeth assembly, dipper bottom plate, dipper side wall, and teeth. **Figure 7** shows the representative stress profile. The stresses on the dipper-teeth assembly vary with time. The maximum stress values vary from 151 MPa to 282 MPa. These stress values are higher than the lower limits of yield strengths for low, medium, and high carbon steel (**Table 5**). Permanent damage to the dipper components is possible, if the steel used has lower yield strength. The stress contour maps are used to identify the high and lower stress regions for fatigue fracture modeling and analysis.

http://dx.doi.org/10.5772/65333

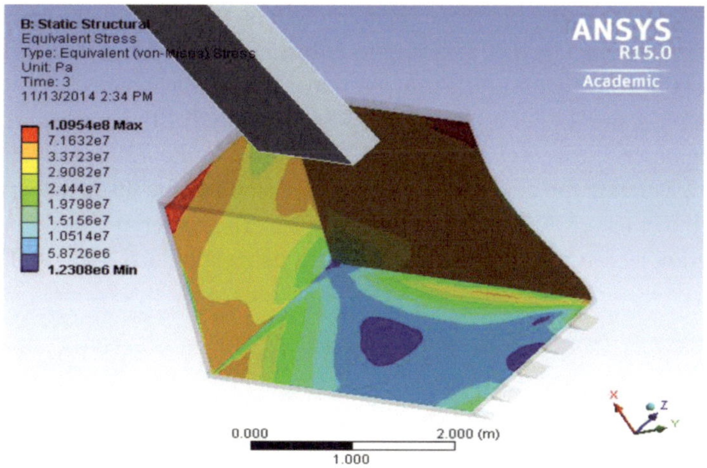

Figure 7. Equivalent stress (von Mises) profile of dipper.

Steel	Yield strength [MPa]
Low carbon steel	140–2400
Medium carbon steel	245–1740
High carbon steel	375–3340

Table 5. Yield strengths of steel [23].

8. Fatigue failure modeling of dipper components

Rope shovel excavation is cyclic in nature. The stresses on the front-end assembly vary continuously during the duty cycle [24]. This variation, combined with the material flaws, can initiate fatigue cracks in shovel components. Environmental factors (e.g., freezing temperatures and corrosive materials) affect metal toughness. Fatigue crack may grow rapidly to undesirable lengths under cyclic loading conditions. There are three modes for metal fatigue failure: (i) mode-I (crack opening), (ii) mode-II (in-plane shear or crack opening), and mode-III (out-of-plane shear or crack twist). Metal failure can also be a result of mixed-mode fatigue. Mode-I fatigue research has dominated the fatigue analysis and life-expectancy field. Three common fatigue failure analysis approaches are typically used, including the stress life, strain life, and fracture mechanics. Each approach has its own application with overlapping boundaries. The fracture mechanics approach is used to estimate a crack's propagation life. For this approach, the initial crack lengths are either assumed or known (welds, known defects, porosities, and cracks found during nondestructive testing). Fracture mechanics principles and theories are applied to estimate crack-propagation rates and, thus, crack-propagation lives.

A typical crack-growth curve has three distinct regions: crack initiation, crack propagation, and rapid increase in crack growth leading to failure. Fatigue life may, however, occur for a longer period of time during the crack-propagation phase as the majority of crack time is spent during this phase. A number of models are available to predict the crack-propagation phase (the middle region on the curve). Paris' law [25], defined by Eq. (32), is the most commonly used method to estimate the crack propagation. The slope of the linear region of the fatigue curve defines the crack-growth rate with every cycle. The material constants (C, m) can be found for different metals in literature or computed using standard tests (ASTM E647):

$$\frac{da}{dN} = C\left(\Delta K\right) \tag{32}$$

K is the stress intensity factor (SIF). According to the linear elastic fracture mechanics (LEFM) theory, computation of stress intensity factor (SIF) at the crack tip is necessary to predict the crack growth. SIFs for many simple geometries and loading situations are available in published literature [26–29]. For complex geometries and stress loading conditions, numerical methods are used to compute the reliable SIF values. The most common methods for calculating SIF are the J-integral and energy release rate. Finite element techniques evaluate the SIF using the energy release rate method. Many commercial software packages, such as ANSYS Workbench (R15), have options for calculating these parameters. In these techniques, energy release is estimated around the crack tip nodes in close loops, in the form of contours. For 2D cases, the node at the crack tip forms the first contour, while for 3D cases, all nodes forming the crack front determine the first contour. The shape, length, and depth of a crack determine the crack life at a specific location.

For fracture modeling of the dipper, representative cracks can be induced in the virtual shovel. **Figure 8** illustrates the location and geometry of an elliptical crack in the dipper's bottom plate. The crack is induced in the high stress region and is along the stress lines. This representative crack is 3 in. long and is 1 in. deep. A localized reference system defines the geometry of this crack. The crack plane lies on this XZ plane, while the width of crack is along the Y direction of this plane. The crack grows along the X and Z directions. A number of parameters control the crack failure modeling, including crack geometry, shape, orientation, and stress environment. The crack definition is explained in **Figure 9**. This crack front is divided into a number of segments, and six circular contours (also divided into segments) are generated around the crack front. All these divisions represent the node locations for the finite element model. The J-integral values are computed for every contour along the crack front. These values are used to compute the SIFs at the crack tip and for life estimations.

The crack size is the most critical aspect of fatigue crack modeling and dipper life estimation. The SIFs are highly dependent upon the crack size (length and depth). The relationship between SIF and crack length is nonlinear. Estimating the variation of SIFs with size is the most important and critical aspect for life estimation. The experimentation is performed for multiple size cracks. ANSYS R15 software is used to compute the J-integrals around the predefined crack tips. As explained in **Figure 9**, contours are generated around the crack front to represent the closed paths for J-integral. There are six contours around every modeled crack as illustrated in **Figure 10**. A very fine mesh size is generated around the crack tip, and J-integrals are computed for all these contours. The first contour is very close to the tip and may represent erroneous results. Therefore, the J-integral values for the first contour are ignored to ensure accuracy [30].

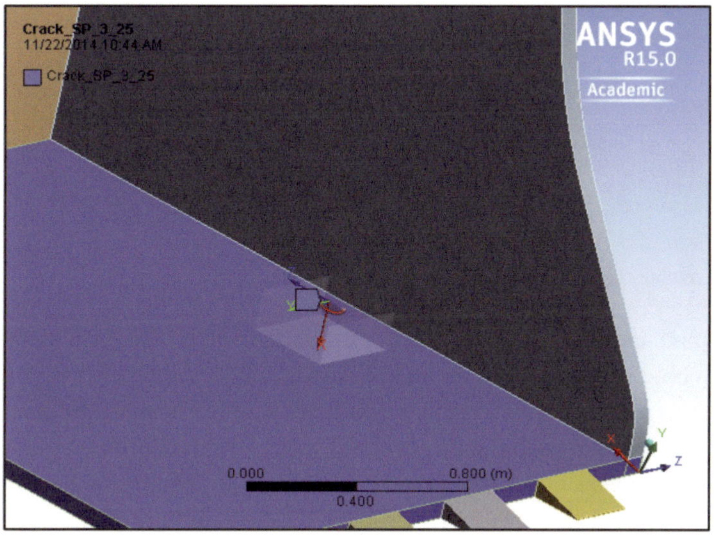

Figure 8. Elliptical crack at the side of the dipper bottom plate.

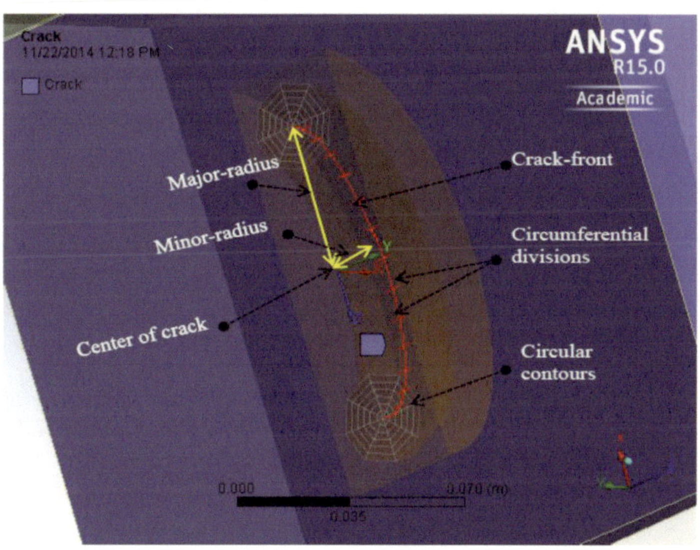

Figure 9. Crack definition in ANSYS for J-integral computations.

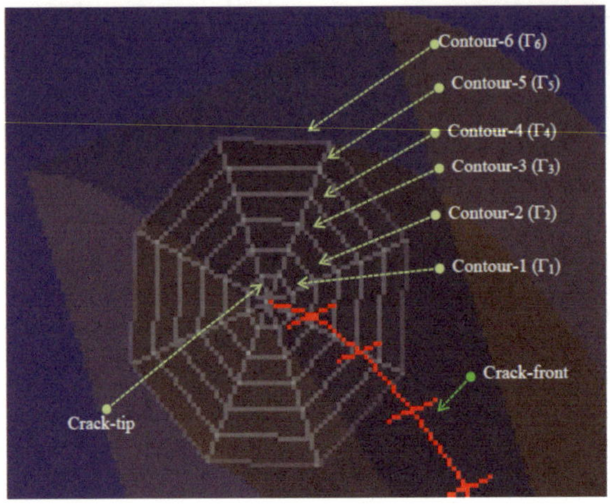

Figure 10. ANSYS contours for J-integral computations.

SIFs are computed, using plane stress conditions for all the contours, and an average value of five contours (contour 2–6) is used for further fatigue analysis. Crack lengths are increased from smaller to larger crack sizes at selected locations, and SIFs are computed for each crack size. The results are used to generate the crack-growth curves and for life expectancy of dipper components. The SIF variation curves for the bottom-plate crack tip are obtained through a least square regression and curve fitting process, Eq. (33), and are plotted in **Figure 11**. The crack is in a high stress region, and, thus, the SIF is very high. Further, the SIFs show a steep increase with crack size. It is expected that the cracks at this location will propagate rapidly:

$$SIF = 4.98E-04a^3 - 8.75E-02a^2 + 5.14a + 2.96E+01 \tag{33}$$

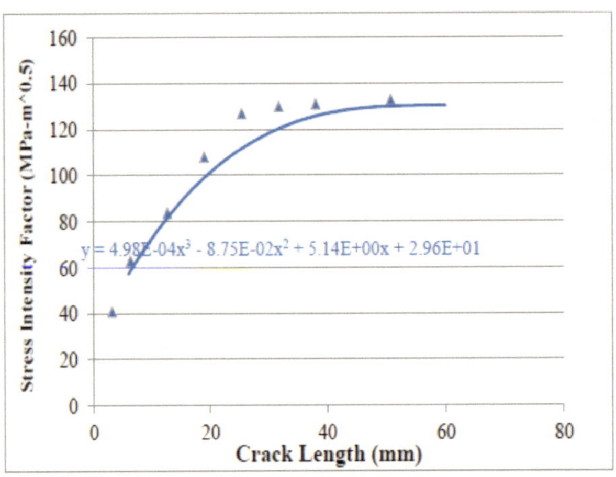

Figure 11. SIFs for the bottom-plate crack tip.

9. Remaining life expectancy of dipper components

Fatigue life is modeled by integrating the Paris' law [25] in Eq. (34). The equation has three input parameters (c, m, and ΔK). The two material constants are computed following standard laboratory procedures. For common materials, the values for these variables can also be found in literature [31, 32]. The "c" values are between 3 and 4, and some common values for "m" are available in literature [33]. For this research, the material constants are taken from research conducted by Yin et al. [21, 22]. They estimated the crack growth for the shovel boom cracks and measured the material constants ("c" and "m") following the ASTM standard E1820:

$$N_f = \int_{a_i}^{a_f} \frac{da}{m(\Delta K)^c} \tag{34}$$

It is also assumed that the steel properties for the dipper and teeth are similar to that for the boom. With these parameters, Eq. (34) becomes Eq. (35). As the computations become complex, it is numerically solved using Gauss-Legendre quadrature in MATLAB. The outputs from Eq. (35) include number of cycles (N_f) for a crack to propagate from an initial length (a_i) to a final length (a_f). The number of cycles is converted to number of days assuming that one digging cycle of shovel is equivalent to one fatigue cycle:

$$N_f = \int_{a_i}^{a_f} \frac{da}{5.89^{-12}(\Delta \mathrm{K})^{3.27}} \tag{35}$$

Following the Palmgren-Miner's rule [34] for equivalent damage, the total number of fatigue cycles per day is equal to the digging cycles of the shovel per day. The total number of cycles for a shovel per day is counted using the cycle time and the operational efficiency. The shovel digging cycle is assumed to be 3 s for this research. However, a typical complete excavation cycle time for P&H 4100XPC is about 30 s. The 3 s cycle time is consistent with the numerical simulation results. Using this cycle time and assuming a 95% shovel operational efficiency, the total number of digging cycles for shovel is calculated as 2730 cycles per day. This assumption is very close to field observations [22] where researchers counted 2880 cycles per day for a cable shovel working continuously over a period of 2 weeks. For this research, a middle-ground value of 2800 cycles per day is assumed to convert the cycles to days.

The remaining useful life for the cracked components can be estimated, with knowledge of critical crack lengths for dipper material. The critical crack length is the length of the crack at which the material at the crack tip starts behaving like a plastic material, and the crack propagation becomes very rapid. It is represented as the boundary between the second and third zones for a fatigue crack. The critical lengths for metals are generally measured using laboratory fatigue toughness tests following the standard procedures. A critical length limit may also be implemented based on field operating conditions or using the crack-growth curves.

The crack-growth curve for the dipper bottom plate is shown in **Figure 12**. It is observed that the crack-propagation rates become very high after a certain crack length. A critical length[1] of 100 mm is set for this crack. As illustrated in this figure, the estimated life for a 50 mm bottom-plate crack is 38 days. However, once the crack grows to 75 mm, the remaining life is only 16 days for the same crack.

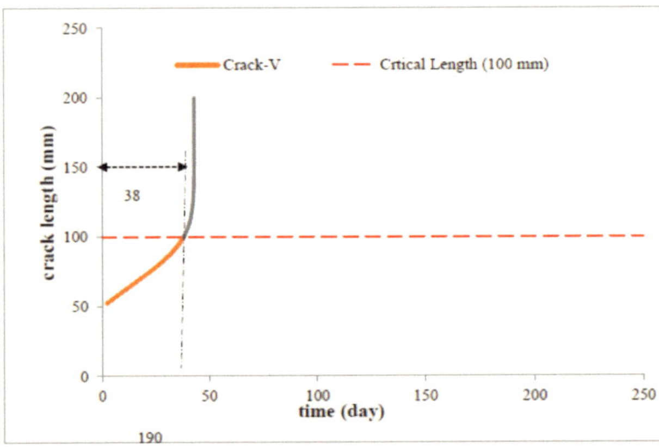

Figure 12. Critical crack length and remaining life expectancy for an initial length of 50 mm.

10. Significant contributions

The research underlying this chapter is a pioneering effort for understanding electric rope shovel dipper stress analysis using dynamic resistive forces. The research results contribute to the existing body of knowledge on health and longevity of shovel dipper-teeth assembly. It advances shovel reliability, maintainability, and availability, which influence surface mining productivity. Previous research generally ignored the dynamic forces due to the weight of the dipper and payload. Given the size of current large shovels with 100+ tons per pass, these forces must be accounted for in any meaningful and comprehensive dynamic model. The research models provide detailed information on the forces and torques for all joints and links within the shovel front-end assembly. This research is also the first attempt to model the fatigue life of the shovel dipper-teeth assembly. It lays a foundation for understanding dipper fatigue failure resulting from high stress intensity, crack initiation, and propagation for rope shovel front-end assembly. Research shows that dipper-related breakdowns are among the highest for shovel excavation downtimes [4]. The current maintenance practice for the shovel front-end assembly is based on experience rather than science. This chapter lays a foundation for

[1] Crack length in this text (and in literature) is always referred as half of the total length of crack. A critical length of 100 mm would be 200 mm total length of the crack.

the scientific modeling and understanding of the shovel dipper-teeth stress and fatigue failure studies. The life expectancy of the shovel components should help reduce the operating costs of shovel excavation. Overall, the study contributes to the health and longevity of the large rope shovels by providing a more scientific basis to the subject matter and should be helpful in the design and development of the next generation of excavators. The results of these models can be used to design and build the next generation of shovel dippers for the surface mining industry and advance frontiers and knowledge in shovel dipper stress and fatigue failure modeling.

11. Conclusions

Estimating electric rope shovel health and longevity is a complicated task and requires a thorough understanding of the shovel digging process. Thus far, the shovel front-end assembly repair model is based on experience and judgment rather than science. Shovel kinematic and dynamic models provide a scientific basis for shovel repair and fatigue failure modeling. The shovel dynamic model requires a good estimation of shovel digging forces. Current rope shovels have large dipper capacities, and their digging resistive forces can be significant. Shovel payload is a dynamic and significant contributor of these digging forces. The dynamic forces result in stress loading of shovel front-end components. The maximum stress values can be as high as 282 MPa and can be higher than the lower yield strength limits for low, medium, and high carbon steel. Material flaws, high stresses, and other environmental factors can initiate cracks on the dipper. Under severe stress loading conditions, these cracks can propagate to critical lengths in no time. The estimated life for a 50 mm bottom-plate crack was found to be 38 days. However, once the crack grows to 75 mm, the remaining life can be as low as 16 days only. This chapter lays a foundation for the fatigue failure analysis of dipper-teeth assembly. The virtual prototype is a scaled and simplified model. It is recommended that the work be extended for real full-scale virtual prototype for the actual steel materials used.

Nomenclature

C_o number of impacts to sink a cylindrical tip in a standardized test by 10 cm

ϱ the angle that the rupture surface makes with the horizontal

β tool cutting angle for Zelenin model

d tool working depth, depth of dipper into the bench

w width of tool

z coefficient for teeth configuration in the Zelenin model [11]

w dipper width

da/dN crack-growth rate per cycle as defined by Paris' law

a	crack length
N_f	number of cycles to failure
K	stress intensity factor
C, m	material constants for Paris' law
a_i, a_f	initial and final/critical known/assumed crack length
$C(\Theta,\bar{\Theta})$	generalized Coriolis and centripetal torque
$D(\Theta)$	generalized inertia matrix
$G(\Theta)$	generalized gravity torque
m_1, m_2	mass of crowd arm and dipper, respectively
αi	rotation of coordinate frame with respect to i−1 frame
β	constant inclination of link 3 from link 2 (inclination of X_4 from X_3)
θ_i	rotation of *i*th coordinate frame
θ	inclination of coordinate frame {4′} from coordinate frame 3
a_i	length of the *i*th link
a_1	crowd-arm length from pivot to connection point between arm and dipper (length of 1st link)
a_2	length between dipper tip and connect point of arm and dipper (length of second)
s_i, c_i	$\sin\theta_i$ and $\cos\theta_i$, respectively
di	offset distance of the gravity left in link i
F_n, F_t	normal and tangential cutting resistive forces on dipper tip
${}^{i+1}T_i$	transformation matrix to transfer the coordinates from i to i+1
${}^{i+1}R_i$	rotational matrix, extracted from ${}^{i+1}T_i$
L(x)	straight-line function representing the bench face

Author details

Muhammad Azeem Raza[1] and Samuel Frimpong[2*]

*Address all correspondence to: frimpong@mst.edu

1 University of Engineering and Technology, Lahore, Pakistan

2 Missouri University of Science and Technology, Rolla, USA

References

[1] Caterpillar. *Electric Rope Shovels*. 2012 [cited 2012]; Available from: https://mining.cat.com/products/surface-mining/electric-rope-shovels.

[2] P&H. *P&H Mining Equipment: Electric Rope Shovels*. 2011 [cited 2011 March 10, 2011]; Available from: http://www.phmining.com/en/PHMining/Mining-Equipment/Electric-Shovels/4100XPC.htm.

[3] Roy, S.K., M.M. Bhattacharyya, and V.N.A. Naikan, *Maintainability and Reliability Analysis of a Fleet of Shovels*. Transactions of The Institution of Mining & Metallurgy, 2001: A163–A171.

[4] S. K. Roy, M. M. Bhattacharyya & V. N. A. Naikan (2001) Maintainability and reliability analysis of a fleet of shovels, Mining Technology, 110:3, 163-171, DOI: 10.1179/ mnt.2001.110.3.163

[5] Lipsett, M.G. and Y.R. Moghaddam, *Bifurcations, Instabilities and Degradation in Geomaterials, SSGG*, in *Modeling Excavator-Soil Interaction*; W. Richard, M.Alsaleh, LabuzJoe, Editor. 2011, Springer.com: 347–366.

[6] Hemami, A., *An Approximation of the Weight of the Loaded Material During the Scooping Operation of a Mechanical Loader*. Transactions of the Canadian Society of Mechanical Engineering, 1994. 18(3): 191–205.

[7] Hemami, A., S. Goulet, and M. Aubertin, *Resistance of Particulate Media to Excavation: Application to Bucket Loading*. International Journal of Surface Mining, Reclamation and Environment, 1994. 8(3): 125–129.

[8] Awuah-Offei, K., S. Frimpong, and H. Askari-Nasab, *Formation Excavation Resistance Modeling for Shovel Dippers*. International Journal of Mining and Mineral Engineering, 2009. 1(2): 127–146.

[9] Takahashi, H., M. Hasegawa, and E. Nakano, *Analysis on the Resistive Forces Acting on the Bucket of a Load-Haul-Dump Machine and a Wheel Loader in the Scooping Task*. Advanced Robotics, 1999. 13(2): p. 97–114.

[10] Balovnev, V., I, *New Methods for Calculating Resistance to Cutting of Soil*. 1983: Amerind, New Delhi, India.

[11] Zelenin, A.N., V.I. Balovnev, and I.P. Kerov, *Machines for Moving the Earth*. 1985: Amerind, New Delhi, India.

[12] Koivo, A.J., *Fundamentals for Control of Robotic Manipulators*. 1989: Jhon Wiley & Sons, Inc. New York, NY, USA

[13] Craig, J.J., Introduction to Robotics: Mechanics and Control. 2nd ed. 1996: Prentice Hall, Upper Saddle River, New Jersey, USA.

[14] Frimpong, S., Y. Hu, and K. Awuah-Offei, *Mechanics of Cable Shovel-formation Interactions in Surface Mining Excavations*. Journal of Terramechanics, Elsevier Ltd., New York, NY 2005: p. 15–33.

[15] Wu, H., *Modeling and Simulation of Electric Mining Shovels*, in *Department of Mining and Metallurgical Engineering*. 1995, McGill University, Montreal, Quebec, Canada.

[16] Frimpong, S., Y. Hu and H.I. Inyang. Dynamic Modeling of Hydraulic Shovel Excavators for Geomaterials. International Journal of Geomechanics. 2008 Vol. 8(1), ASCE, Reston, VA: 2–10.

[17] Hendricks, C. and M. Scoble. *Post-Blast Evaluation Through Shovel Performance Monitoring*. In *Proc. of the Conference on Explosive and Blasting Technique*. 1990, © ISEE, Orlado, FL.

[18] Sommer H.J., *POLYGEOM.m: Geomery of a Planar Polygon*. 2011.

[19] P&H. *4100XPC Electric Mining Shovel*. 2012 [Sep, 30, 2012]; Available from: http://www.minepro.com/MinePro/Literature/Spec/4100XPC-AC.pdf.

[20] Coulomb, C.A., *Essai Sur Une Application Des REgles Des Maximis et Minimis a Quelques Problems de Statique Relatifs a l'architechture*. Academic Royale des Sciences: Memories de Mathmatique et de Physique, presentes a l'Acaemie Royale des Sciences, par Divers Savants, Paris, 1776. 7: 343–382.

[21] Yin, Y., et al., *Fatigue life prediction of heavy mining equipment. Part 2: Behaviour of corner crack in steel welded box section and remaining fatigue life determination*. Journal of Constructional Steel Research, 2008. 64(1): 62–71.

[22] Yin, Y., et al., *Fatigue life prediction of heavy mining equipment. Part 1: Fatigue load assessment and crack growth rate tests*. Journal of Constructional Steel Research, 2007. 63(11): p. 1494–1505.

[23] matweb. *Material Property Dataseet*. 2014 [cited 2014 12/2/2014]; Available from: http://www.matweb.com/search/QuickText.aspx?SearchText=carbon%20steel.

[24] Frimpong, S. and Y. Hu, *Intelligent Cable Shovel Excavation Modeling and Simulation*. International Journal of Geomechanics (c) ASCE, 2008. 8(1): p. 2–10.

[25] Paris, W.G. and F. Erdogan, *A Critical Analysis of Crack Propagation Laws*. Trans. of ASME. Journal of Basic Engineering, 1963. D85: p. 528–534.

[26] Loakimidis, N.I. and P.S. Theocaris, *The Numerical Evaluation of a class of generalized stress intensity factors by use of the Lobatto-Jacobi numerical integration rule*. International Journal of Fracture, 1978. 14: p. 469–484.

[27] Raju, I.S. and J.C. Newman, Jr., *Three Dimensional Fine-Element Analysis of Finite-Thickness Fracture Specimens*. NASA Langely Research left, United States., 1997: p. 42.

[28] Sih, G.C., *Handbook of stress-intensity factors*. 1973: Lehigh Univ., Institute of Fracture and Solid Mechanics. Bethlehem, Pennsylvania, USA.

[29] Tada, H., Paris, P.C. and G.R. Irwin, *Stress Analysis of Cracks Handbook*. 1973: Del Research Corp., Hellertown, PA.

[30] ANSYS, *ANSYS Workbench Help Documentation*. 2014, ANSYS Inc. Southpointe 275 Technology Drive Canonsburg, PA, USA.

[31] Chapra, S.C. and R.P. Canale, *Numerical Methods for Engineers: with PC Applications*. 1985: McGraw-Hill, NY.

[32] Rolfe, S.T. and J.M. Barson, *Fracture and Fatigue Control in Structures*. 1977: Prentice-Hall, Englewood, NJ.

[33] Throop, J.F. and G.A. Miller, *Optimum Fatigue Crack Resistance Acheivment of High Fatigue Resistance in Metals and Alloys, ASTM STP 467*. 1970: American Society for Testing and Materials, Philadelphia.

[34] Miner, M.A., *Cummulative Damage in Fatigue*. Journal of Applied Mechanics, 1945. 67.

Chapter 6

Lagrangian Model-Based Fault Diagnosis in a PVTOL

César Martínez Torres,
Luis Humberto Rodríguez Alfaro,
Efrain Alcorta Garcia, Gerardo Romero Galvan and
David Lara

Additional information is available at the end of the chapter

http://dx.doi.org/10.5772/66395

Abstract

A Lagrangian formalism is used to model a PVTOL in order to obtain an aircraft model. The Euler-Lagrange model of the PVTOL is used to develop an algorithm for fault diagnosis. Diagnosis implies the detection, isolation and identification of a fault. The considered approach is based on the knowledge of a system model as well as the model of the possible faults. The idea is to use non-linear decoupling approach to derivate a set of subsystems, each related to a specific fault or a set of faults. An observer-based residual generation is designed for each subsystem, this structure allows the fault detection and isolation stage, for fault identification a kind of approximated inversion algorithm to meet the different diagnostic levels. The results are obtained taking advantage of the structure given by the Euler-Lagrange modelling of the PVTOL as well as from recent results related to observer design and fault identification.

Keywords: fault diagnosis, Lagrangian systems, non-linear systems, observers, fault isolation

1. Introduction

Nowadays, the unmanned aerial vehicles (UAVs) represent a big boom in the electronic industry, thanks to their versatility and largely due to the falling cost of the electronic parts and the UAV by itself. UAV is a kind of an aerial vehicle that is able to take off vertically, such as helicopters and some special airplanes, and it is represented by the planar vertical take-off landing (PVTOL) aircraft model. Note that PVTOL aircraft models represent more than only UAV systems. Reliability requirements in aerial vehicles bring the necessity of a fault detection and isolation schema. In general, they are non-linear systems, and so a non-linear inspired strategy for the detection and isolation of faults could also be used. An idea consists in taking advantage of the structure given by the Lagrangian model of a PVTOL in order to develop an algorithm for the detection, isolation and identification of faults.

Many research studies dealing with the fault detection and isolation (FDI) problem have been already published, most of them deals with linear systems, see for instance Refs. [1–3]. On the other hand, for non-linear systems, some solutions exist, based on the inherited characteristics, see Refs. [4, 5] for more details. The most common approach used for FDI is the hardware redundancy; however, this approach normally represents an increment in weight and economical cost of the aircraft. In order to avoid this problem, some mathematical relations could be used, the simplest way is to compare two or more internal signals, having as goal to create a residue, which, in fact will be zero if the system is working normally and different from zero if not. In order to create such relations it is common to exploit some intrinsic characteristics of the systems. See for instance Ref. [6]. Diagnosis for the PVTOL system has been considered previously using a Hamiltonian formalism [7].

A Lagrangian formalism is used to model a PVTOL in order to obtain an aircraft model. The Euler-Lagrange model of the PVTOL is used to develop an algorithm for fault diagnosis. Diagnosis implies the detection, isolation and identification of a fault. The considered approach is based on the knowledge of a system model as well as the model of the possible faults. The idea is to use non-linear decoupling approach to derivate a set of subsystems, each related to a specific fault or a set of faults. An observer-based residual generation is designed for each subsystem. Detection and isolation of faults can be reached at this stage, for fault identification a kind of approximated inversion algorithm to meet the different diagnostic levels. The results are obtained taking advantage of the structure given by the Euler-Lagrange modelling of the PVTOL as well as from recent results related to observer design and fault identification.

Fault diagnosis algorithms can be developed for a more or less general Euler-Lagrange model of a system, which, in fact, also include a PVTOL system. Fault diagnosis includes detection, isolation and identification of faults. In order to meet a diagnosis task, an observer-based residual generator is designed in order to determine whether a fault is present. A decoupling approach is used in order to guarantee also a fault isolation task. As discussed, both steps could be systematically developed for the considered system model. Further, fault isolation is approached using a kind of approximated system inversion to develop approximated fault estimation through dynamic inversion of the corresponding residual equation. The schema is shown using a specific example of a PVTOL. As presented in the results, the proposed approach can be used effectively for the diagnosis of a PVTOL system.

2. Lagrangian modelling

There is a huge amount of literature about Lagrange's equations of movement, however, see for example Ref. [8]. The structure of a *Planar-Vertical-Take-Off* and *Landing* (PVTOL) is represented in **Figure 1**.

The absolute linear position to the PVTOL is defined in the inertial frame $x-y-z$ axes with two generalized coordinates $\xi^T \triangleq [y \quad z]$. One additional generalized coordinate, the angular position,

http://dx.doi.org/10.5772/66395

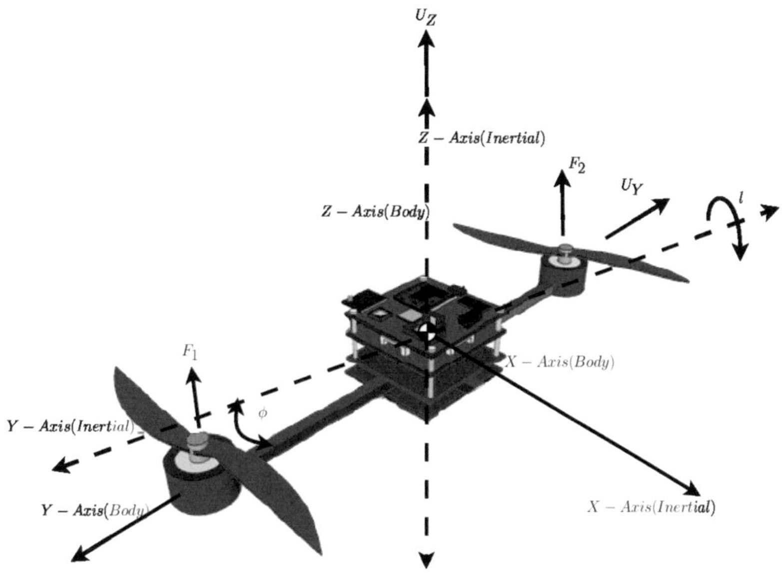

Figure 1. Schema of the PVTOL system.

is defined in the inertial frame. Note that the pitch angle θ, i.e. the rotation angle around the y-axis, and yaw angle Ψ, i.e. the rotation of the PVTOL around the z-axis, are zero. The only angular movement is the roll angle ϕ, i.e. the rotation around the x-axis.

$$\xi = \begin{bmatrix} y \\ z \end{bmatrix}, \quad \eta = \phi, \quad q = \begin{bmatrix} y \\ z \\ \phi \end{bmatrix} \tag{1}$$

The origin of the body frame (also the origin of the inertial frame) is the centre of mass of the PVTOL system. The PVTOL is assumed to have a symmetric structure with the two arms aligned with the body x-axis. The inertia is represented by J_x.

The Lagrangian is defined as the sum of kinetic energy minus the potential energy (E_{pot}). In the case of the PVTOL, the kinetic energy consist of two parts, one related to the translational energy (E_{tran}) and the second related to the rotational energy (E_{rot}).

$$\mathcal{L}(q,\dot{q}) = E_{tran} + E_{rot} - E_{pot} \tag{2}$$

The movement equations of Lagrange are given by

$$\frac{d}{dt}\left[\frac{\partial \mathcal{L}(q,\dot{q})}{\partial \dot{q}}\right] - \frac{\partial \mathcal{L}(q,\dot{q})}{\partial q} = \begin{bmatrix} f_y \\ f_z \\ \ell \end{bmatrix} \tag{3}$$

where f_y represent the generalized forces on the y-axis, f_z represent the generalized forces on the z-axis and ℓ is the torque.

For the PVTOL results:

$$E_{\text{tran}} = \frac{1}{2} m [\dot{y} \quad \dot{z}] \begin{bmatrix} \dot{y} \\ \dot{z} \end{bmatrix} \tag{4}$$

$$E_{\text{rot}} = \frac{1}{2} J_x \omega^2 = \frac{1}{2} J_x \dot{\phi}^2 \tag{5}$$

$$E_{\text{pot}} = mgz \tag{6}$$

So that the Lagrangian results

$$\mathcal{L}(q,\dot{q}) = \frac{1}{2} m\dot{y}^2 + \frac{1}{2} m\dot{z}^2 + \frac{1}{2} J_x \dot{\phi}^2 - mgz \tag{7}$$

and the terms

$$\frac{\partial \mathcal{L}}{\partial \dot{q}} = \begin{bmatrix} m\dot{y} \\ m\dot{z} \\ J_x \dot{\phi} \end{bmatrix} \tag{8}$$

$$\frac{d}{dt}\left(\frac{\partial \mathcal{L}}{\partial \dot{q}}\right) = \begin{bmatrix} m\ddot{y} \\ m\ddot{z} \\ J_x \dot{\omega} \end{bmatrix} = \begin{bmatrix} m\ddot{y} \\ m\ddot{z} \\ J_x \ddot{\phi} \end{bmatrix} \tag{9}$$

$$\frac{\partial \mathcal{L}}{\partial q} = \begin{bmatrix} 0 \\ mg \\ 0 \end{bmatrix} \tag{10}$$

The generalised forces (in the inertial frame) are given by

$$f_y = \cos(\phi) U_y - \sin(\phi) U_z \tag{11}$$

$$f_z = \sin(\phi) U_y + \cos(\phi) U_z \tag{12}$$

where U_z represents the total thrust force (the sum of the forces of each rotor), acting on the z-axis of the body frame. U_y corresponds to the side forces on the y-axis of the body frame. The moment acting on the rolling angle is given by ℓ.

The movement equations are given by

$$\begin{bmatrix} m & 0 & 0 \\ 0 & m & 0 \\ 0 & 0 & J_x \end{bmatrix} \begin{bmatrix} \ddot{y} \\ \ddot{z} \\ \ddot{\phi} \end{bmatrix} + \begin{bmatrix} 0 \\ mg \\ 0 \end{bmatrix} = \begin{bmatrix} \cos(\phi) & -\sin(\phi) & 0 \\ \sin(\phi) & \cos(\phi) & 0 \\ 0 & 0 & 1 \end{bmatrix} \begin{bmatrix} U_y \\ U_z \\ \ell \end{bmatrix} \tag{13}$$

3. Diagnosis approach

Fault diagnosis aim to detect the fault occurrence in the functional units of the system, as well as to classify the different faults and to determine the type, magnitude and cause of faults, which leads to undesired behaviour of the whole system. The fault diagnosis can be achieved by

http://dx.doi.org/10.5772/66395

hardware redundancy or software redundancy also called analytical redundancy. One technique of fault diagnosis is the model-based fault diagnosis, which employs software redundancy.

In the model-based fault diagnosis technique, the system behaviour is online reconstructed by a mathematical model, which is implemented in the software form. In this scheme, the system model run in parallel to the system and both of them are driven by the same control inputs. Thus, in the fault-free case, reconstructed system variables by the system model follow the corresponding real system variables and show a derivation in the faulty case.

A comparison of the measured system variables with their estimates by the system model is called residual. Thus, a residual signal includes the fault effect, and ideally if the residual signal is different from zero, then a fault has occurred otherwise the system is fault free. The residual generation process is carried out in two stages, first, the system outputs have to be estimated, then, the difference between those signals and the signal coming from sensors is computed [9].

Figure 2 shows the general scheme for residual generation using a model-based fault diagnosis technique.

In this contribution, a fault diagnosis for systems with model Euler-Lagrange is presented. A model-based fault diagnosis technique with analytical redundancy is used to obtain a residual generation.

Consider a dynamic system without faults described by the following Euler-Lagrange equations

$$\begin{aligned} \frac{d}{dt}\left[\frac{\partial L(q,\dot{q})}{\partial \dot{q}}\right] - \frac{\partial L(q,\dot{q})}{\partial q} &= \tau, \\ y_o &= q, \end{aligned} \tag{14}$$

where $\tau \in \mathbf{R}^n$ is the vector of generalized forces, $q \in \mathbf{R}^n$ is the vector of generalized coordinates, L is the Lagrangian and y is the vector output.

In this work, additive faults in control input and sensor are considered. The Euler-Lagrange model of the faulty system is defined as

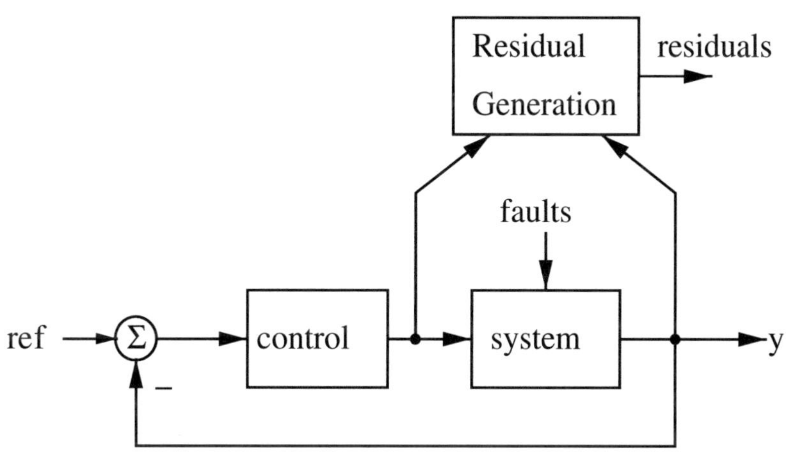

Figure 2. General scheme for residual generation.

$$\frac{d}{dt}\left[\frac{\partial L(q,\dot{q})}{\partial \dot{q}}\right] - \frac{\partial L(q,\dot{q})}{\partial q} = (\tau + QF_a), \qquad (15)$$
$$y_f = (q + NF_s),$$

where $F_a \in \mathbf{R}^n$ is the vector of control input faults, $Q \in \mathbf{R}^{n\times n}$ is a constant matrix, $F_s \in \mathbf{R}^n$ is the vector of sensor faults and $N \in \mathbf{R}^{n\times n}$ is a constant matrix.

Assumption 1. *Consider an Euler-Lagrange system with faults described by Eq. (15) and the system behaviour is on line reconstructed by the Euler-Lagrange system without faults Eq. (14), then the faults presented in the system Eq. (15) can be detected by the residual generator*

$$r(t) = y_f(t) - y_o(t) \qquad (16)$$

4. Application results

The method presented in the previous section is applied in a PVTOL. Only additive faults are taken into account, the faults could affect sensors (y, z and ϕ) and the control inputs (u_z, u_y and ℓ), the faulty case is restricted to one fault at a time, meaning that it is assured that if a fault appears, it is impossible that another fault occurs. Once the fault has occurred, it still presents until the end of the simulation.

The faulty PVTOL system is defined as

$$\begin{bmatrix} m\cos(\phi) & m\sin(\phi) & 0 \\ -m\sin(\phi) & m\cos(\phi) & 0 \\ 0 & 0 & J_x \end{bmatrix}\begin{bmatrix} \ddot{y} \\ \ddot{z} \\ \ddot{\phi} \end{bmatrix} + \begin{bmatrix} mg\sin(\phi) \\ mg\cos(\phi) \\ 0 \end{bmatrix} = \begin{bmatrix} U_y \\ U_z \\ \ell \end{bmatrix} + \begin{bmatrix} 1 & 0 & 0 \\ 0 & 1 & 0 \\ 0 & 0 & 1 \end{bmatrix}\begin{bmatrix} f_{uy} \\ f_{uz} \\ f_\ell \end{bmatrix}, \qquad (17)$$

$$y_o = \begin{bmatrix} y \\ z \\ \phi \end{bmatrix} + \begin{bmatrix} 1 & 0 & 0 \\ 0 & 1 & 0 \\ 0 & 0 & 1 \end{bmatrix}\begin{bmatrix} f_{sy} \\ f_{sz} \\ f_{s\phi} \end{bmatrix}, \qquad (18)$$

where f_{uy} is the fault in the control input U_y, f_{uz} is the fault in the control input U_z, f_ℓ is the fault in the control input ℓ, f_{sz} is the fault in the sensor of position in the vertical movement, f_{sy} is the fault in the sensor of position in the horizontal movement and $f_{s\phi}$ is the fault in the sensor of angle ϕ.

Among the different detection methods available in the literature [2], the threshold-based approach is one of the most common thanks to its simplicity and accuracy. The principle of this approach is based on the idea that the parameters of the system (e.g. mass and dimensions) could vary because of the measurement or estimation errors. Those data are used to determine a threshold, which is computed by varying the internal parameters of the system in a certain ± percentage. This is carried out in order to avoid false alarms caused by the difference between the mathematical model and the real system.

http://dx.doi.org/10.5772/66395

The internal parameters of the PVTOL that may vary are the mass (m) and the inertia (J_x), in order to cover the worst case scenario, both parameters vary at the same time +10 and −10%. As a result, the thresholds are fixed as depicted in **Figure 3**.

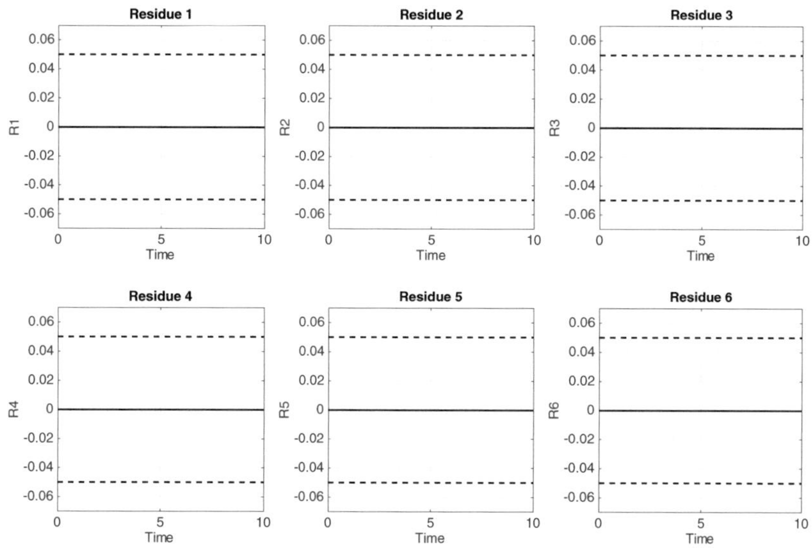

Figure 3. Amplitudes of the detection thresholds. - -, detection threshold; —, residues.

As explained in the beginning of this section only additive faults are taken into account, since the controller is designed to stabilize the system in hover flight, the fault amplitude is defined as a percentage of the initial value for sensors and a percentage of the maximum amplitude of the input control. This percentage is fixed ±10% for sensors and ±5% for control inputs. The faults are triggered 7 seconds after the beginning of the simulation and it is persistent until the end.

In order to detect the fault, six different residues are computed, for this, it is assumed that the entire state is available, according to the previous section as follows:

$$R_1 = y_f - y \quad (19)$$

$$R_2 = z_f - z \quad (20)$$

$$R_3 = \phi_f - \phi \quad (21)$$

$$R_4 = \dot{y}_f - \dot{y} \quad (22)$$

$$R_5 = \dot{z}_f - \dot{z} \quad (23)$$

$$R_6 = \dot{\phi}_f - \dot{\phi} \quad (24)$$

where the suffix$_f$ denotes the signal coming from the sensor.

4.1. Sensor faults

The sensor faults considered in this work affect the vertical measurement (z), the horizontal (y) and the inclination angle (ϕ), as explained before the amplitudes of the faults are 0.2 m, 0.1 m and 0.2°, respectively. The controller is designed to decouple the sensor faults, as a result each fault is independent of the others and by consequence all of them are detectable and isolable, thanks to their different fault signatures. **Figures 4–6** depict the sensor faults.

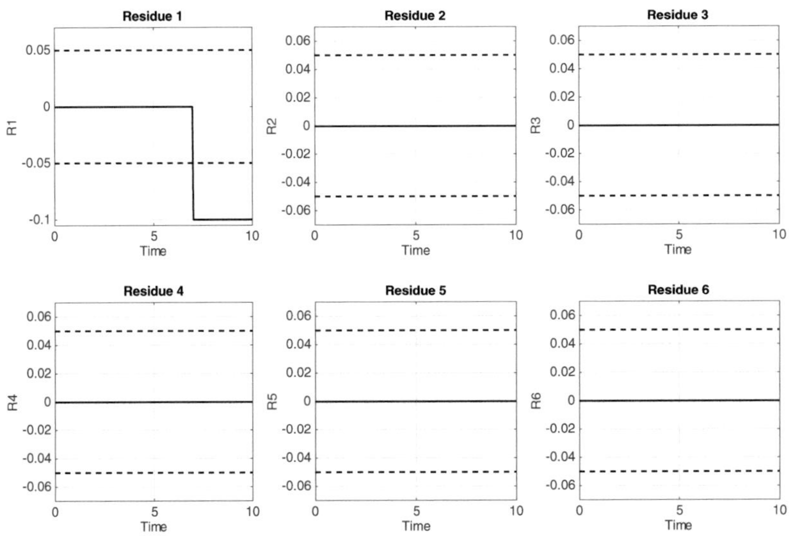

Figure 4. Fault affecting y sensor. - -, detection threshold; —, residues.

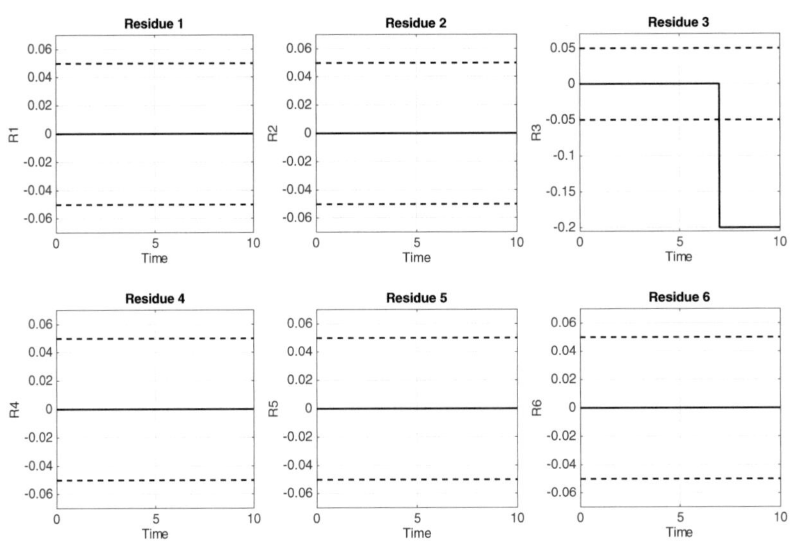

Figure 5. Fault affecting z sensor. - -, detection threshold; —, residues.

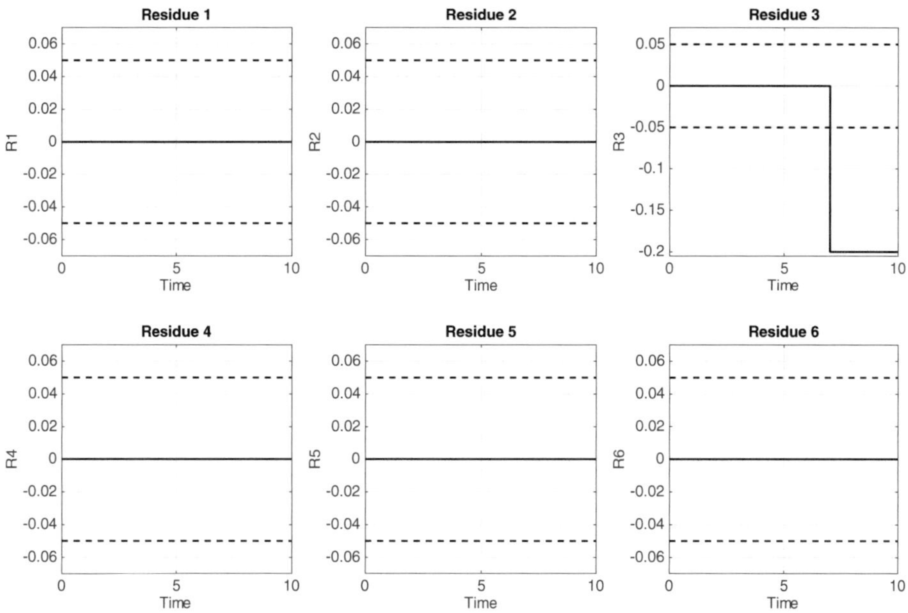

Figure 6. Fault affecting ϕ sensor. - -, detection threshold; —, residues.

Once the residue exceeds the detection threshold, the fault is considered detected and it will be isolable if and only if the fault signature is different among the others. As expected, and thanks to the controller design every fault affecting the sensors is isolable. The fault signatures are presented in **Table 1**. *X* means that the residue exceeds the threshold; *O* means that even if the residue is affected, it does not surpass the threshold and by consequence this residue is not triggered.

Fault	R1	R2	R3	R4	R5	R6
Sensor y	*X*	*O*	*O*	*O*	*O*	*O*
Sensor z	*O*	*X*	*O*	*O*	*O*	*O*
Sensor ϕ	*O*	*O*	*X*	*O*	*O*	*O*

Table 1. Fault signatures of sensors.

4.2. Control inputs faults

The fault amplitudes of the control inputs are fixed by obtaining the 5% of the maximum size of them during an unfaulty simulation, after this processes the amplitudes are fixed to 0.5, 0.425 and 0.008 for u_y, u_z and ℓ, respectively. Thanks to the controller design, the faults affecting the control inputs are detectable and isolable. **Figures 7–9** shows that the detection threshold is exceeded once a fault occurs, by consequence, fault detection is accomplished. **Table 2** depicts the fault signatures, it is straightforward to see that they are all different, this behavior confirms that every single fault is detected and isolated.

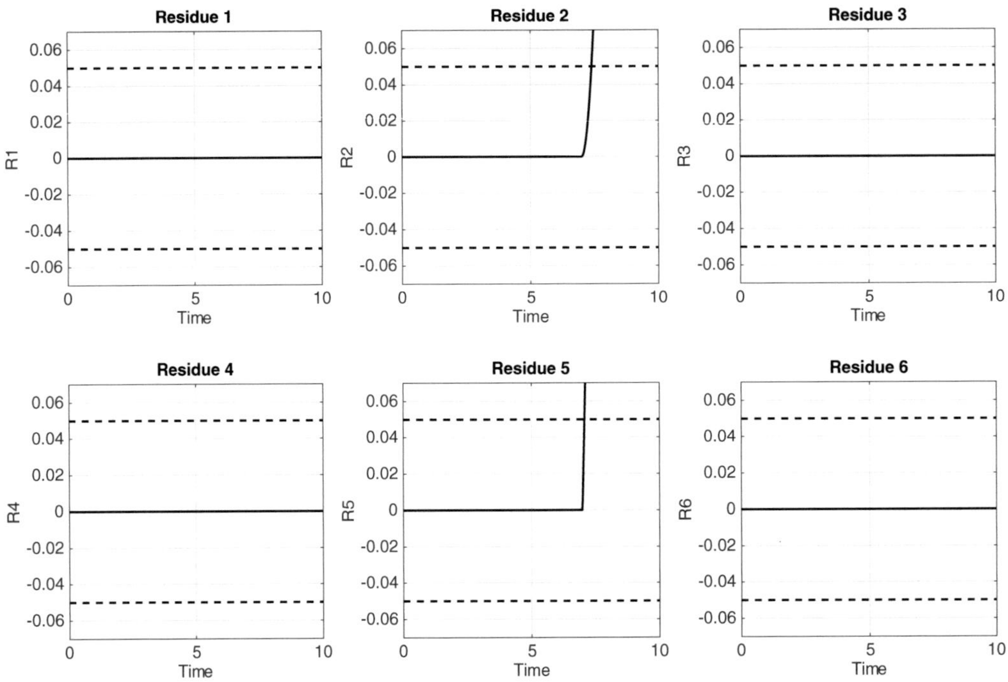

Figure 7. Fault affecting u_z. - -, detection threshold; —, residues.

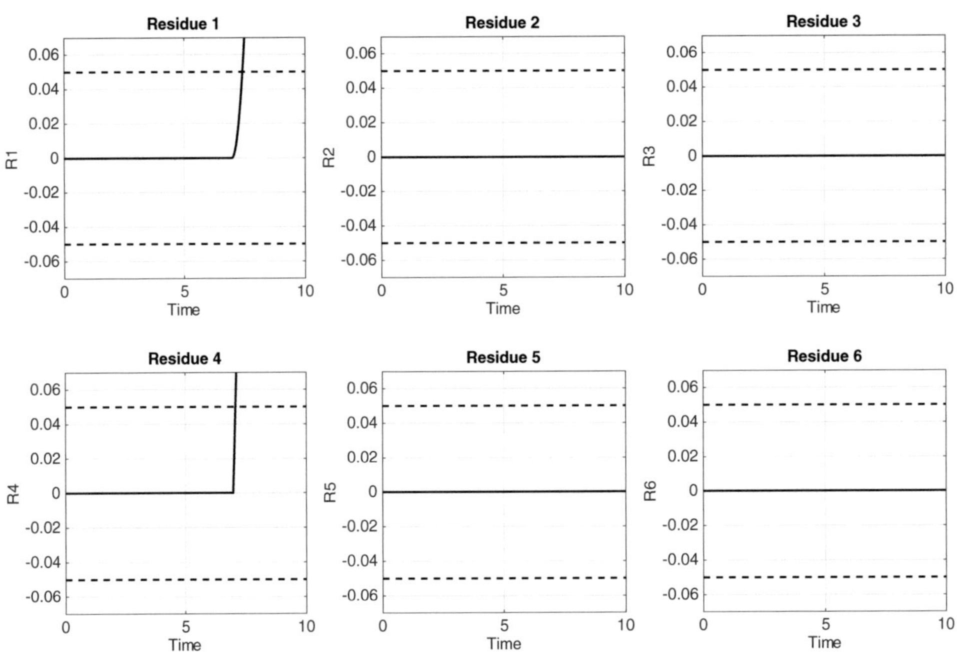

Figure 8. Fault affecting u_y. - -, detection threshold; —, residues.

http://dx.doi.org/10.5772/66395

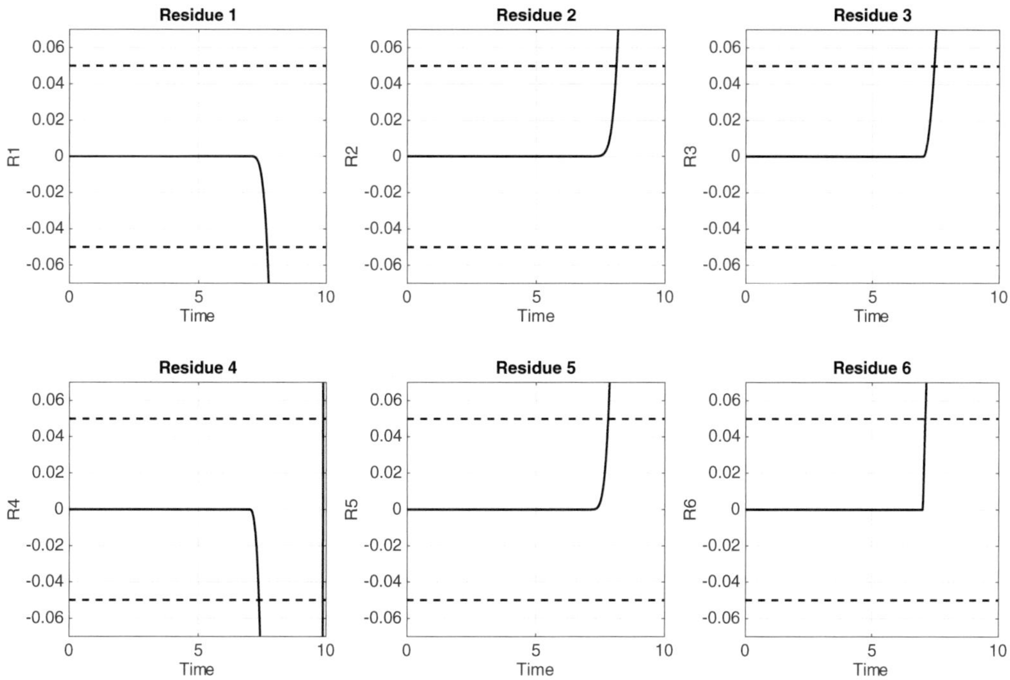

Figure 9. Fault affecting ℓ. - -, detection threshold; —, residues.

Fault	R1	R2	R3	R4	R5	R6
Control input u_z	*O*	*X*	*O*	*O*	*X*	*O*
Control input u_y	*X*	*O*	*O*	*X*	*O*	*O*
Control input ℓ	*X*	*X*	*X*	*X*	*X*	*X*

Table 2. Fault signatures of control inputs.

5. Conclusion

This work presents a fault detection and isolation approach applied to Lagrangian systems.

Every fault is detectable and isolable as it can be seen in **Tables 1** and **2**, this result is obtained thanks to the special design of the state feedback controller, by consequence the faults affecting the sensors are easily isolable, and the residues affected during a control input fault are those related to the measures affecting the states, for instance a fault affecting u_y triggers the residues related to the y and $\dot{y}$ measures; on the other hand, the faults affecting the control input ℓ trigger all the residues, this is because the direct relation between this control input and the ϕ measure, as could be seen in Eq. (13) besides the ϕ measure appears in the other states, as a result every residue is triggered; however, this signature is unique, by consequence this fault is considered detected and isolated.

Author details

César Martínez Torres[1], Luis Humberto Rodríguez Alfaro[2], Efrain Alcorta Garcia[2*], Gerardo Romero Galvan[3] and David Lara[3]

*Address all correspondence to: efrain.alcortagr@uanl.edu.mx

1 Universidad de las Américas Puebla, Mexico

2 Universidad Autónoma de Nuevo León, Mexico

3 Universidad Autónoma de Tamaulipas, Mexico

References

[1] J. Chen and R. J. Patton. Robust model based fault diagnosis for dynamic systems. Kluwer Academic Publishers Group, Springer US, 1999.

[2] M. Blanke, M. Kinnaert, and J. Lunze M. Staroswiecki. *Diagnosis and fault-tolerant control.* Springer, Berlin Germany, 2nd edition, 2006.

[3] S. X. Ding. Model-based fault diagnosis techniques. Springer-Verlag London, 2nd edition, 2013.

[4] E. Alcorta-Garcia and P. M. Frank. Deterministic nonlinear observer-based approaches to fault diagnosis: A survey. *Control Engineering Practice,* 5(5):663–670, May 1997.

[5] C. DePersis and A. Isidori. A geometric approach to nonlinear fault detection and isolation. *IEEE Transactions on Automatic Control,* 46(6):853–865, June 2001.

[6] C. Martinez Torres, L. Lavigne, F. Cazaurang, E. Alcorta Garcia, and D. Diaz Romero. Fault tolerant control of a three tank system: A flatness based approach. In *Control and Fault-Tolerant Systems (SysTol), 2013 Conference on,* pp. 529–534, Oct 2013.

[7] L. H. Rodriguez Alfaro, E. Alcorta Garcia, D. Lara, and G. Romero. A Hamiltonian approach to fault isolation in a planar vertical take-off and landing aircraft model. *International Journal of Applied Mathematics and Computer Science,* 25(1):65–76, 2015.

[8] B. C. Fabien. Analytical system design. Springer US, 2009.

[9] S. X. Ding. Model-based fault diagnosis techniques. Springer-Verlag Berlin Heidelberg, 2008.

Chapter 7

Fuzzy Logic and *S*-Lagrangian Dynamics of Living Systems: Theory of Homeostasis

Uziel Sandler and Lev Tsitolovsky

Additional information is available at the end of the chapter

http://dx.doi.org/10.5772/66473

Abstract

A key peculiarity of living organisms is their ability to actively counteract degradation in a changing environment or being injured by using homeostatic protection. In this chapter, we propose a dynamic theory of homeostasis based on a recently proposed generalized Lagrangian approach (*S*-Lagrangian). Following the discovery of homeostasis W. Cannon, we assume that homeostasis results from the tendency of the organisms to decrease the stress and avoid death. We show that the universality of homeostasis is a consequence of analytical properties of the *S*-Lagrangian, while peculiarities of the biochemical and physiological mechanisms of homeostasis determine phenomenological parameters of the Lagrangian. We show that plausible assumptions about *S*-Lagrangian features lead to good agreement between theoretical descriptions and observed homeostatic behavior.

Keywords: Homeostasis, *S*-Lagrangian, Dynamics, living systems, stress

1. Introduction

A primary difference between living creatures and non-living things is the capacity for reproduction. However, if one considers only individual life rather than the existence of species, the major paradox is that living things actively counteract degradation in a continuously changing environment or being injured through homeostatic protection. By homeostasis, we refer to the ability of living organisms to maintain viability and stability of physiological functions in a changing external environment. The system remains alive as a consequence of homeostasis maintaining system integrity in the presence of perturbing influences. Cessation of homeostasis leads to inevitable death. In living systems, the relationship between cause and effect is paradoxical: organisms are characterized by poorly predictable motility, which is supposedly managed by their internal motives. Homeostatic motivation transforms an object into a subject by virtue of its own behavior. Thus, the mystery of arbitrary actions may be disclosed by exploring homeostasis [1]. It should be noted that homeostasis may evidently produce both maintenance of life and the will to act [2].

Although homeostasis is present in all living systems and relates to large numbers of different biochemical and physiological mechanisms, it reveals amazingly similar features and behavior. Such universality is not unique in the physical world. For example, physical systems, from crystals to large biomolecules, demonstrate universal behavior near critical points in spite of considerable differences in its structures and intermolecular interactions. This occurs due to the critical behavior of the systems being determined by the analytical properties of free energy near critical points, while the peculiarities of system structure and intermolecular interactions are "hidden" within the phenomenological parameters of the free energy.

We assume that the universality of homeostasis is a consequence of the analytical properties of the *S-Lagrangian*, which determines the dynamic equation associated with homeostasis, while peculiarities of the biochemical and physiological mechanisms determine phenomenological parameters of the Lagrangian. We show in Section 2 that plausible assumptions about *S-Lagrangian* properties lead to good agreement between theoretical descriptions and observed homeostatic features.

2. Biological background

2.1. Homeostasis levels

Living beings actively oppose their degradation in continuously changing environments by means of homeostasis [3] that supports the intrinsic bodily constants within acceptable limits. Maintenance of individual life requires evaluating and regulating its inner state. Homeostatic regularities can be traced to the level of particular cellular parameters, cells, in general, physiological systems of an organism, and an organism as a whole. In this study, we primarily focus on homeostasis of neurons and the nervous system. A cell, as a body, manifests complete homeostasis. This occurs not only to maintain biological constants but also to regulate physiological functions and motivational behavior. The behaving animal is sensitive to single neuronal spikes and even to their temporal patterning [4]. Moreover, a neuronal spike can serve as a tool of reaction for the whole animal [1]. Individual neurons act in concert to govern behavior [5].

At first glance, homeostatic mechanisms are not complicated. In theoretical research, the problem is often evaluated by the introduction of positive- and negative-feedback loops between the sensor and the metabolic flaw (e.g., [6, 7]). Attempts to model homeostatic regulation consider only simple homeostasis, with regulation of each variable described by the introduction of specific individual controllers. However, when homeostatic protection begins to work against a permanent environmental factor or severe injure, these mechanisms become ineffective and living systems utilize indirect paths to assign optimal parameters, depending on the situation.

Homeostatic function depends on sensors, which register deviations from the norm. Appearance of a metabolic flaw triggers the homeostatic device to compensate for the shortage.

http://dx.doi.org/10.5772/66473

However, homeostatic resources may not be sufficient to restore disturbed functions. In these cases, living systems may try to change the environment, requiring the environment to be included in the interaction.[1]

The status of the internal environment is not sustainable for all life. Conditions remain stable only at intervals of time as compared to environmental variability. At these intervals, homeostasis counteracts weak disorders in the system and recovers initial conditions (*direct regulation*). Over time, adapting to strong external influences enables life to modify its parameters (*indirect regulation*). If the value of a deviated parameter is not restored, the organism may be able to maintain it by restructuring the optimum of other parameters. For example, stabilization of neuronal activity can be achieved by configuring both synapse efficiency and cell autonomous homeostasis [8]. Homeostasis readjusts to save some supreme quality criterion that distinguishes the living from the nonliving. The living entity keeps track of a special criterion the degree of remoteness from its destruction. This criterion determines the intensity of homeostatic protection. However, damage may reach such an extent that homeostasis is unable to overcome the irreversible destruction of the living system.

The nature of the general sensor for damage-recovery viability is unclear, though there are options that are significant to the survival of cells and the whole organism. These include energy (ATP level), excitability, intracellular pH levels, and concentration of certain proteins (caspases, cytokines, or antioxidants). These cannot be disregarded by the highest sensors, which could lead to death. For example, a supreme neuronal sensor might be excitability [1].

2.2. Protection generates action

Misalignment of homeostasis leads to damage, the increase in the activity, and leads to further aggravation of injury. As a rule, the response of neurons is proportional to the coming stimulation. However, superfluous stimulation and neuronal injury are intimately connected (excitotoxicity) [9]. Thus, the injured neurons generate spikes.

It should be noted that extensive damage of nerve tissue reduces excitability and violates its function, while protection temporarily restores excitability. Therefore, there is a region of the paradoxical states of excitable tissues, where excitation is reduced due to damage, but irreversible deterioration of the tissue has not yet occurred. In such a case, inhibition (or decreased excitation) counteracts the damage, paradoxically recovers the normal excitability, and promotes the generation of action (parabiosis, in accordance with N.E. Vedensky) [10]. Properties of homeostatic protection make it tempting to consider homeostasis as a driving force that induces actions directed against actual or anticipated damage. However, in cases where the damage cannot be completely compensated for by available resources, metabolic problems may be solved through actions directed at the environment.

To outside observers, the resulting behavior will resemble the emergence of motivation, will to live, and be match with conscious decision. The optimal state corresponds to such conditions that do not threaten the lives and do not evoke attempts to change structure and functioning of

[1]We do not consider this complicated form of homeostasis in this study; however, our approach is extendable to this case, as well.

the brain. A minimum of homeostatic load will serve as a criterion for this state. Joint behavior of a huge variety of such systems generates to complex forms of awareness. The neural tissue concentrates within itself the ability to evaluate its own state and endeavor to survive. A single cell, neurons in particular, can live, learn, want, suffer, delight, and try to survive [1]. Exertion of neuronal metabolism, leading to protection and goal-directed behavior, is rather appropriate for the description of conscious actions. Purposeful behavior corresponds to conscious decision and resembles a kind of generalized "pursuit of life." To the outside observer, this is reminiscent of intentional action and a manifestation of will.

2.3. Emergence of the feeling of a death threat

The essence of subjective feeling, goals, and will is still the amazing mysteries. The establishment of the theory of systems regulating optimal constants of their own state gives hope for understanding the problem of subjectivity, as homeostasis is the key tool that supports the system alive. The emergence of self as a state separated from the external environment is probably a direct consequence of the vitality of living systems. Alive system should assess its own vitality, and the phenomenon of maintaining its life is impossible to distinguish from instrumental actions. Probably, life appears along with the ability of an individual to evaluate its own integrity and health, and the homeostatic protection is a material manifestation of the pursuit of life.

Maintenance of vital activity resembles a manifestation of the mystical "vital force," which prevents disorder and violates the laws of thermodynamics. Homeostatic activities are so rational that their discoverer, W. Cannon, described them as "Wisdom of the body."

The appearance of self-dissatisfaction plays a crucial role in triggering homeostatic protection, especially in the emergence of aware decisions. Nevertheless, it is difficult to provide a formal definition for the subjective feelings of discomfort that coincide with the appearance of damage. The assessment of general parameters is qualitative and is guided by the "injure-repair" scale. Living systems somehow regulate the avoidance of injury and the aspiration to life as it shifts toward death or life, that is, behaves as an object possessing minimum awareness. This mysterious variable may not be a function of the state and should depend on the previous history of the system, since homeostasis, as well as behavior, improves after exercise [1].

We have no possibility of determining how a neuron evaluates its own state, but we know that injury decreases positive feelings, while protection decreases distress. In any case, the approach of death increases cellular efforts to operate. A living system reacts to damage as if it is having a negative sensation. Homeostasis entails a relationship between physiology and mind. The problems of consciousness and the problem of life self-maintenance are inseparable. It is likely that the origin of life necessarily leads to the emergence of consciousness.

2.4. Homeostatic regulation

Theoretically, there are two explanations for homeostatic operation:

1. Rigid mechanistic programs that evaluate all options for possible injures.
2. Spontaneous relaxation, which minimizes injury.

If it is algorithmically predetermined by the Genetics, the body needs to recognize its own current state and select a genetically pre-tuned course of recovery. However, the number of possible optima can be as high as the number of non-lethal states of external environment and this would create invalid load on the genome. Besides, genetic defects often have limited impact on the relevant functional paths, since homeostasis is capable of compensating for many such defects [11, 12]. Therefore, it is reasonable to assume that spontaneous recovery to a sustained state is the main mechanism of homeostasis.

In general, a living system is open and its dynamics is irreversible. Living beings are somehow able to evaluate their remoteness or closeness to death. While this is beyond doubt, we cannot specify the exact mechanism of evaluation. Movement of a living being within the space of its parameters should minimize this global parameter, that is, proximity to death.

3. Theory of homeostasis

3.1. Dynamics equations of homeostasis

Consider a living organism, whose state is described by n variables, $q = \{q_1, \ldots, q_n\}$. These variables can describe both behavioral and physiological or neurophysiological features and we consider them as coordinates of the abstract state space of the system.[2] As we have mentioned in Section 1.4, living organisms are somehow able to evaluate their level of discomfort or stress (see Ref. [1] for comprehensive discussion), so we consider this feature as additional scalar variable, $S_,$ and will call S as *stress-index* (*S-index*). It is a typical phenomenological variable, which cannot be directly measured,[3] but it should be emphasized that although S corresponds to the "feeling" quantity, it is an objective feature of the living beings [1].

In experiments with living organisms, many parameters that influence on the system's behavior are out of control, which leads to considerable deviations in *numerical* values of the experimental results. It means that small differences in the values of the experimental data became insignificant and the state of a system should be described by a *domain* of points rather than a *single* point in the state space. This kind of uncertainty does not have stochastic nature and L. Zadeh has introduced for its notion of the *fuzzy sets* [13] and theory of *possibility* [14–16].

We assume that the dynamics of the living systems satisfies causality principle in the form (see Ref. [1] for details):

- "If, at the time $t + dt$, the system is located in the vicinity of the point x, then at the previous time t, the system could be near the point $x' \approx x - \dot{x}' dt$, or near the point $x'' \approx x - \dot{x}'' dt$, or near the point $x''' \approx x - \dot{x}''' dt$, or ..., and so on, for all possible values of the velocity $\dot{x}$."

[2] We assume that the state space has trivial local topology, which means that any inner point of any small domain in the space belongs to the space as well.

[3] The phenomenological variables, which cannot be directly measured, are widely used in physics, for example "mechanical action" of the physical systems, order parameter of the superfluid phase transition, and so on.

where $x = \{q, S\}$. Since velocities $\{\dot{q}, \dot{S}\}$ cannot be precisely obtained, we describe them by the function $Pos(\dot{q}, \dot{S}; q, S, t)$,[4] which indicates *possibility* that the system has velocities $\{\dot{q}, \dot{S}\}$ near the point $\{q, S\}$ at the time t. The most possible velocities satisfy

$$Pos(\dot{q}, \dot{S}; q, S, t) = 1 \tag{1}$$

and only this case will be considered in this chapter.

It has been shown in Ref. [17] that if a system's evolution satisfies the causality principle, the system's state space has trivial local topology, and if state can be described by a compact fuzzy set, then the most possible system's trajectories $\{q(t), S(t)\}$ satisfy the generalized Lagrangian-like equations

$$\frac{d}{dt}\frac{\partial L}{\partial \dot{q}_i} - \frac{\partial L}{\partial q_i} = \frac{\partial L}{\partial S}\frac{\partial L}{\partial \dot{q}_i}, \tag{2a}$$

$$\frac{dS}{dt} = L(\dot{q}, q, S, t), \tag{2b}$$

where $L(\dot{q}, q, S, t)$ is the solution of Eq. (1) with respect to $\dot{S}$. (We will call $L(\dot{q}, s, S, t)$ as "*most possible S-Lagrangian*" or *S*-Lagrangian for short. The equations of motion (2a) and (2b) are more general than the common Lagrangian equations. Since these equations can describe the dynamics of sets, they can be differential inclusion instead differential equations. The second extension is *dependence of the Lagrangian on S-variable*[5] (*S*-Lagrangian). In this case, the Lagrangian equations of motion acquire a non-zero right side, proportional to the derivative of the *S*-Lagrangian with respect to *S*. It has been shown in Ref. [17] that the equations of motion with *S*-Lagrangian lost time reversibility, the energy and momentum are not conserved even in closed systems. Note that *S*-Lagrangian is not an invariant under the addition of a function which is a total derivative with respect to time.[6] It should be emphasized that the derivation of these equations in Ref. [17] does not depend on any specific properties of the system or its Lagrangian. This means that Eqs. (2a) and (2b) give a reasonable method of applying the Lagrangian approach to non-physical systems. So, we believe that the dynamics of homeostasis can be described by Eqs. (2a) and (2b) with appropriate choice of the *S*-Lagrangian $L(\dot{q}, q, S, t)$.

Attempting to decrease stress and proximity to death is a basic feature of the living organisms. It is important that this feature exists already on a single-cell level (see Ref. [1] for comprehensive discussion). Deviation of the system's parameters from their ground values leads to increasing discomfort and the organisms try to decrease discomfort by generating the protection mechanisms. These mechanisms, in turn, generate the system's activity (see Section 1.3 or

---

[4]It should be emphasized that the function $Pos(\dot{q}, \dot{S}; q, S, t)$ cannot be identified with any probability density $\rho(\dot{q}, \dot{S}; q, S, t)$, because it has different mathematical features. Actually, $Pos(\dot{q}, \dot{S}; q, S, t)$ is a *function*, while $\rho(\dot{q}, \dot{S}; q, S, t)$ is a *functional* [17].

[5]In the classical mechanics, *S-variable* is nothing more than common mechanical action.

[6]In the classical mechanics, *S-variable* is nothing more than common mechanical action.

[1]), which can be described by time derivatives of the variables, $\dot{\boldsymbol{q}}$. Following the discovery of homeostasis W. Cannon [3], we assume that homeostasis results from a tendency of the organisms to decrease the stress and avoid death and that the dynamics of the stress is determined by competition between damage and the protection mechanisms. So we write

$$\frac{dS}{dt} = L(\dot{\boldsymbol{q}},\boldsymbol{q},S,t) = -P(\dot{\boldsymbol{q}},\boldsymbol{q},S) + I(\boldsymbol{q},S,t), \tag{3}$$

where function $I(\boldsymbol{s},S,t)$ describes increasing of stress by deviation of the system's parameters, while $P(\dot{\boldsymbol{q}},\boldsymbol{q},S)$ corresponds to decreasing of stress by the protection mechanisms.

Experimental observations of homeostatic behavior (see Ref. [1] and references there) show that functions $I(\boldsymbol{q},S,t)$ and $P(\dot{\boldsymbol{q}},\boldsymbol{q},S)$ should satisfy the following:

i. Deviation of the system's variables from the ground states corresponds to injury or damage, even if *S-index* does not have the time to change.

ii. If stress is high, the same perturbation of the variables can strongly increase *S-index*, than its increasing at low levels of stress.

iii. Protection is reinforced by moderate stress, but if stress is very high, the protection mechanism becomes less effective.

Below, we consider time intervals, which is much shorter than the time of relevant changes in environmental conditions, so that we can neglect time dependence in Eq. (3) and write

$$L(\dot{\boldsymbol{q}},\boldsymbol{q},S) = -P(\dot{\boldsymbol{q}},\boldsymbol{q},S) + I(\boldsymbol{q},S), \tag{4}$$

and will call $I(\boldsymbol{q},S)$ as *Injure* and $P(\dot{\boldsymbol{q}},\boldsymbol{q},S)$ as *Protection* for short.

By using Eq. (4), we rewrite Eqs. (2a) and (2b) as

$$-\frac{d}{dt}\frac{\partial P}{\partial \dot{q}_i} + \frac{\partial}{\partial q_i}[P{-}I] = \frac{\partial P}{\partial \dot{q}_i}\frac{\partial}{\partial S}[P{-}I] \tag{5a}$$

$$\frac{dS}{dt} = -P(S,\dot{\boldsymbol{q}},\boldsymbol{q}) + I(\boldsymbol{q},S). \tag{5b}$$

Equations (5a) and (5b) are the main *dynamic equations of homeostasis*. It should be noted that S-*index*

$$S = S_0 + \int_0^t [-P(S(t'),\dot{\boldsymbol{q}}(t'),\boldsymbol{q}(t')) + I(\boldsymbol{q}(t'),S(t'))]dt' \tag{6}$$

is not function of a state but depends on the system's history.

For small-to-moderate activity, we can expand $P(\dot{\boldsymbol{q}},\boldsymbol{q},S)$ with respect to $\dot{\boldsymbol{q}}$. We have[7]:

---

[7] Summating on the repeated indices (Einstein summation) is assumed.

$$P \simeq A(\boldsymbol{x},\xi,S) + a_i(\boldsymbol{x},\xi,S)\,\dot{\xi}_i + \frac{1}{2} m_{ij}(\boldsymbol{x},\xi,S)\dot{\boldsymbol{x}}_j\dot{\boldsymbol{x}}_i, \tag{7}$$

where we designate by the Latin symbol: $\boldsymbol{x}$ the variables with zero linear terms in Eq. (7) and by the Greek symbol: ξ the variables with non-zero linear terms[8] and keep in Eq. (7) only the terms with lowest order on $\dot{\xi}$ and $\dot{\boldsymbol{x}}$. For reasons that will be clarified later, we will refer to $\boldsymbol{x}$ as *stationary variables (C-variables)* and ξ as *running variables (R-variables).*

The term $A(\boldsymbol{x},\xi,S)$ corresponds to short-term compensation of stress (e.g., by immediate releasing of the endorphins ("endogenous morphine"), which are quickly produced in natural response to pain [1]). The other terms correspond to long-term protection by generating the activity.[9] In the last terms, matrix m_{ij} determines character rates of changing of the variables $\boldsymbol{x}$: small m_{ii} corresponds to the fast-changing variables, while large m_{jj} corresponds to the slow-changing ones. The function $\boldsymbol{a}(\boldsymbol{x},\xi,S)$ determines the behavior of the *R-variables* (see page 13)).

Therefore, Eqs. (5a) and (5b) take the form:

$$m_{ij}\ddot{\boldsymbol{x}}_j + \left(W\frac{\partial m_{ij}}{\partial S} - \frac{\partial W}{\partial S} m_{ij}\right)\dot{\boldsymbol{x}}_j = -\frac{\partial}{\partial \boldsymbol{x}_i}(W - a_j\dot{\xi}_j), \tag{8a}$$

$$\Omega_{ij}^{-1}\dot{\xi}_j + \frac{\partial a_i}{\partial \boldsymbol{x}_j}\dot{\boldsymbol{x}}_j = \frac{\partial W}{\partial S} a_i - W\frac{\partial a_i}{\partial S} - \frac{\partial W}{\partial \xi_i}, \tag{8b}$$

$$\frac{dS}{dt} = -\frac{1}{2} m_{ij}\dot{\boldsymbol{x}}_j\dot{\boldsymbol{x}}_i + \left(W(\boldsymbol{x},\xi,S) - a_j\dot{\xi}_j\right). \tag{8c}$$

where we designated

$$W(\boldsymbol{x},\xi,S) = I(\boldsymbol{x},\xi,S) - A(\boldsymbol{x},\xi,S). \tag{9}$$

$$\Omega_{ij}^{-1} = \frac{\partial a_i}{\partial \xi_j} - \frac{\partial a_j}{\partial \xi_i} + a_i\frac{\partial a_j}{\partial S} - a_j\frac{\partial a_i}{\partial S}. \tag{10}$$

and in the first approximation with respect to $\dot{\boldsymbol{x}}$ and $\dot{\xi}$ we have omitted in Eqs. (8a) and (8b) the terms that are proportional to $o(\dot{\boldsymbol{x}}_k\dot{\boldsymbol{x}}_j, \dot{\boldsymbol{x}}_k\dot{\xi}_j)$.

Since Ω_{ij}^{-1} is an antisymmetric matrix, $\Omega_{ij}^{-1} = -\Omega_{ji}^{-1}$, Eq. (11b) may include the rotation of *R*-variables in the *{ξ}* subspace. This means that even in the ground state, where *C*-variables possess stationary stable points $\dot{\boldsymbol{x}}_c = 0, \dot{S}_c = 0$, *R*-variables are functions of time (this is why we refer to these variables as *running variables*).

By using Eq. (8b), Eqs. (8a) and (8c) can be rewritten as

[8] In order to ensure that P would increase along with increasing activity, the matrix m_{ij} should be positively defined.

[9] Interestingly, various human activities, for example, aerobic exercise, stimulate the release of endorphins as well [18].

http://dx.doi.org/10.5772/66473

$$m_{ij}\ddot{\boldsymbol{x}}_j + \left[W\frac{\partial m_{ij}}{\partial S} - \frac{\partial W}{\partial S}m_{ij} + \frac{\partial}{\partial \boldsymbol{x}_i}\left(a_l\Omega_{lk}\frac{\partial a_k}{\partial \boldsymbol{x}_j}\right)\right]\dot{\boldsymbol{x}}_j = -\frac{\partial U}{\partial \boldsymbol{x}_i}, \tag{11a}$$

$$\dot{\xi}_j = \Omega_{ij}\left(\frac{\partial W}{\partial S}a_j - W\frac{\partial a_j}{\partial S} - \frac{\partial a_j}{\partial \boldsymbol{x}_k}\dot{\boldsymbol{x}}_k\right), \tag{11b}$$

$$\frac{dS}{dt} = -\frac{1}{2}m_{ij}\dot{\boldsymbol{x}}_j\dot{\boldsymbol{x}}_i + \left(a_i\Omega_{ij}\frac{\partial a_j}{\partial \boldsymbol{x}_k}\right)\dot{\boldsymbol{x}}_k + U(\boldsymbol{x},\xi,S), \tag{11c}$$

where

$$U = W\left(1 + a_j\Omega_{jk}\frac{\partial a_k}{\partial S}\right) + a_j\Omega_{jk}\frac{\partial W}{\partial \xi_k}. \tag{12}$$

Equations (11a) and (11c) represent *dynamic equations of homeostasis* for the systems with temperate activity.

3.2. Behavior near the stable states

In order for the running variables to not disturb the ground state, $S_c = 0$, $\boldsymbol{x}_c = \text{const.}$, we should assume that

$$\frac{\partial}{\partial \boldsymbol{x}_{kc}}a_j(\boldsymbol{x}_c,\xi,S_c) = 0; \ \frac{\partial}{\partial S_c}a_j(\boldsymbol{x}_c,\xi,S_c) = 0; \ \frac{\partial}{\partial \xi_k}W(\boldsymbol{x}_c,\xi,S_c) = 0. \tag{13}$$

(see Eqs. (11a) and (12)).

Stable states of Eqs. (11a) and (11c) are defined by

$$W(\boldsymbol{x}_c,S_c) = \ = 0, \tag{14a}$$

$$\frac{\partial W}{\partial \boldsymbol{x}_{ci}} = 0. \tag{14b}$$

There are two types of solutions for Eqs. (14a) and (14b), which could be called as ground states (*GSSs*) and as local stable states (*LSSs*). At *GSS*, the injure reaches its global minimum $I(\boldsymbol{x}_{c1},S_{c1}) = 0$ that leads to

$$A(\boldsymbol{x}_{c1},S_c) = 0. \tag{15}$$

In order for Eq. (15) to be valid for any set of $\boldsymbol{x}_{c1}$ that satisfy Eq. (14a), the function $A(S_c,\boldsymbol{x}_{c1})$ should be factorized as

$$A(\boldsymbol{x}_{c1},S) = S\Psi(\boldsymbol{x}_{c1},S), \tag{16}$$

given that $\Psi(\boldsymbol{x}_{c1},S)\neq 0$ (see Eq. (22)).

Unlike at *LSS*, where the system remains injured

$$I(\boldsymbol{x}_{c2},S_{c2}) > 0, \tag{17}$$

S-index is non-zero, because

$$A(\boldsymbol{x}_{c2},S_{c2}) > 0 \Rightarrow S_{c2} > 0. \tag{18}$$

This means that near *LSS*, the system is stressed, but its state is stable.

Consider the case where $m_{ij} = m_{ij}(\boldsymbol{x},S)$ and $W = W(\boldsymbol{x},S), \boldsymbol{a} = \boldsymbol{a}(\xi)$. If deviations from the stable state

$$\boldsymbol{y} = \boldsymbol{x}-\boldsymbol{x}_c, \tag{19a}$$

$$w = S-S_c, \tag{19b}$$

are small, we can expand Eqs. (11a) and (11c) with respect to $\boldsymbol{y}$ and w. In the first-order approximation, we obtain[10]

$$\ddot{y}_i + \gamma_c\dot{y}_i = -K_{ij}y_j, \tag{20a}$$

$$\dot{w} = -\gamma_c w. \tag{20b}$$

where

$$\gamma_c = -\frac{\partial W_c}{\partial S_c},$$

$$K_{ij} = m_{ik}^{-1}(\boldsymbol{x}_c,S_c)\frac{\partial^2 W_c}{\partial \boldsymbol{x}_{ck}\partial \boldsymbol{x}_{cj}}.$$

Equations (20a) and (20b) are simple and can be easily solved:

$$y_j = e^{-\gamma_c t/2}\sum_\alpha(Q_{j\alpha}e^{i\omega_\alpha t} + Q_{j\alpha}^* e^{-i\omega_\alpha t}), \tag{21a}$$

$$w = w_0 e^{-\gamma_c t}. \tag{21b}$$

where

$$\omega_\alpha = \sqrt{\lambda_\alpha - \frac{\gamma_c^2}{4}},$$

$w_0,Q_{j\alpha}$ are constants and λ_α are eigenvalues of the matrix, K_{ij}. We see that in order for the stationary state, $\boldsymbol{x}_c$, to be stable, it needs to be

---

[10]The terms that are proportional to w in Eq. (20a) and to y in Eq. (20b) have vanished because of conditions (14a) and (14b).

$$\frac{\partial W_c}{\partial S_c} < 0. \tag{22}$$

Additionally, matrix K_{ij} should be positively defined. The ground states correspond to zero damage and *S-index*, while the disturbed stationary states correspond to the local minimums of $W(\boldsymbol{x},S)$.

Consider the behavior of *R*-variables near the ground state with $\boldsymbol{a} = \boldsymbol{a}(\xi)$. In accordance with Eqs. (11b) and (14a), we have

$$\dot{\xi}_i = -\gamma_c(\boldsymbol{x}_c,S_c)\Omega_{ij}(\xi)\, a_j(\xi), \tag{23}$$

so the behavior of the *R*-variables is determined by the function $\boldsymbol{a}(\xi)$.[11] It is convenient to present ξ in the form $\xi(t) = \xi(t)\boldsymbol{n}(t)$, where $\xi(t)$ and $\boldsymbol{n}(t)$ are the scalar and vector functions, respectively, with $|\boldsymbol{n}| \equiv 1$. Then Eq. (23) takes the form

$$\dot{\xi} = -\gamma_c n_i \Omega_{ij}\, a_j, \tag{24a}$$

$$\dot{n}_i = -\frac{\gamma_c}{\xi}\left(\Omega_{ij}\, a_j - n_i(n_k\Omega_{kj}a_j)\right). \tag{24b}$$

If $\boldsymbol{a} = \varphi(\xi)\xi$, where $\varphi(\xi)$ is a scalar function, these equations are simplified:

$$\dot{\xi} = 0, \tag{25a}$$

$$\dot{n}_i = -\frac{\gamma_c\varphi}{\xi}\Omega_{ij}\, n_j. \tag{25b}$$

Therefore, in this case, $\xi = \xi_0 = \text{const.}$.

In the case of two *R*-variables, we can write ξ as

$$\xi = \xi_0\begin{bmatrix}\cos\varphi\\ \sin\varphi\end{bmatrix}, \tag{26}$$

which implies that $\varphi = \varphi(\cos\phi,\ \sin\phi)$, and Eq. (25b) takes the form

$$\frac{d\phi}{dt} = -\frac{\gamma_c\varphi}{\xi_0}\left(\cos\phi\frac{\partial\varphi}{\partial\sin\phi} - \sin\phi\frac{\partial\varphi}{\partial\cos\phi}\right)^{-1}. \tag{27}$$

Therefore,

[11]Note that because matrix Ω_{ij} is antisymmetric, *R*-variables exist only if there are at least two *R*-variables.

$$\xi_1 = \xi_0 \cos(\phi(t)), \tag{28a}$$

$$\xi_2 = \xi_0 \sin(\phi(t)). \tag{28b}$$

where function $\phi(t)$ should be obtained from Eq. (27).

3.3. Simulation results

For easy visualization of the typical behavior of systems with homeostasis, we consider a system with two *C*-variables and two *R*-variables: $\boldsymbol{x} = \{x_1, x_2\}$, $\xi = \{\xi_1, \xi_2\}$ and $m_{ij} = m_i \delta_{ij}$ with constant $m_1 \ll m_2$, making $\boldsymbol{x}_1$ fast and $\boldsymbol{x}_2$ slow variables. In order to clarify the influence of *C*-variables and *S*-index upon the homeostatic behavior, we choice also $W = W(\boldsymbol{x}, S)$ and a simplest form of $\boldsymbol{a}$

$$a = \begin{bmatrix} a_{01} & 0 \\ 0 & a_{02} \end{bmatrix} \begin{bmatrix} \xi_1 \\ \xi_2 \end{bmatrix}, \tag{29}$$

with constant a_{01}, a_{02}. In this case, Eqs. (11a), (11b) and (11c) are simplified and we have[12]

$$m_i \ddot{x}_i - \frac{\partial W}{\partial S} m_i \dot{x}_i = -\frac{\partial W}{\partial x_i} \tag{30a}$$

$$\dot{\xi}_i = \frac{\partial W}{\partial S} \Omega_{ij} a_j, \tag{30b}$$

$$\frac{dS}{dt} = -\frac{1}{2} m_i \dot{x}_i^2 + W(\boldsymbol{x}, S). \tag{30c}$$

Conditions (i)–(iii) on page 8 allow us to choose the functions $I(\boldsymbol{x}, S)$ and $A(\boldsymbol{x}, S)$ in the form[13]

$$I(\boldsymbol{x}, S) = \Phi_1(S) J(\boldsymbol{x}), \tag{31a}$$

$$A(\boldsymbol{x}, S) = S \Phi_2(S) \Gamma(\boldsymbol{x}), \tag{31b}$$

where $\Phi_1(S)$ and $\Phi_2(S)$ are monotonically increasing and decreasing functions of *S*, respectively, with $\{\Phi_1(0), \Phi_2(0)\} > 0$ and $J(\boldsymbol{x}) \geq 0, \Gamma(\boldsymbol{x}_c) > 0$.[14]

Results of the simulation are shown in **Figures 1** and **2** for the different initial conditions.

[12] There is no summation on *i*.

[13] Generally speaking, both $I(\boldsymbol{x}, S, \xi)$ and $A(\boldsymbol{x}, S, \xi)$ may depend on *R*-variables far from the stable states, but here we have neglected this opportunity.

[14] Simulation shows that the qualitative behavior of $\boldsymbol{x}(t)$, $S(t)$, and $\xi(t)$ weakly depends upon the concrete choice of the functions $\Phi_1(S), \Phi_2(S)$ and $J(\boldsymbol{x})$, $\Gamma(\boldsymbol{x})$ if they satisfy conditions (i)–(iii). For results are shown below we have used $\Phi_1(S) = (1 + bS^k)$, $\Phi_2(S) = (1 + cS^n)^{-1}$, with $b = c = 1$ and $k = n = 2$.

http://dx.doi.org/10.5772/66473

In **Figure 1A**, light injuring of the system causes the main ground state to be slightly disturbed. We see that the fast and slow *C*-variables[15] quickly find their stable points. Injure (**Figure 1A**, row 4) and *S-index* (**Figure 1A**, row 5) approach zero, while the *R*-variables (**Figure 1A**, rows 6 and 4) remain running. Therefore, in this case, homeostasis cares for the injury, fully reduces the stress (*S*-index becomes zero), and returns the system to its main ground state. Interestingly,[16] in spite of the fact that injury and protection can quickly oscillate, *S-index* approaches zero much more smoothly and does not "feel" the quick alteration of the injure parameter (**Figure 1A**, row 4).

In **Figure 1B**, the initial perturbation was somewhat stronger, resulting in the system being unable to return to the main ground state. However, after further trials, homeostasis finds another non-distressing (zero *S-index*) ground state (**Figure 1B**, rows 1 and 2), where injury and distress are vanished, as well (**Figure 1B**, rows 4 and 5).

In **Figure 1C**, the initial perturbation was more stronger, so protection (**Figure 1C**, row 3) cannot fully reduce injury and distress. Nevertheless, homeostasis finds the region of *C*-variables where the system is stable (**Figure 1C**, rows 1 and 2), because protection was able to compensate the injury, but, unlike the previous case, the protection mechanisms should be permanently running. So the system remains damaged and distressed (**Figure 1C**, rows 4 and 5).

Figure 2 shows a situation where the system was heavily injured. We see that protection (**Figure 2A**, row 3) failed to compensate for the injury (**Figure 2A**, row 4) and after short-time damage and stress drastically increasing (**Figure 2A**, rows 4 and 5), *C*-variables leave the life-compatible region (**Figure 2A**, rows 1 and 2) and the system inevitably moves toward death or destruction. We see that crossover to this way can be very sharp. Moreover, in this situation, the behavior of *R*-variables differs considerably from the behavior near the stable states. The system appears to be "crying" in response to the dangerous situation (**Figure 2A**, row 6). Interestingly, a similar situation occurs in the case of an initially strongly stressed system, although the initial injury was small (**Figure 2B**).

It should be emphasized that the decreased protection observed in **Figures 1A** and **B** and **Figure 2** is different. In **Figure 1**, the protection mechanism has done the work and the system returns to its ground state with zero stress and injury, unlike the situation observed in **Figure 2** where protection fails to compensate for the injury and slows down due to the stress level becoming too high.

If a system has a "latent time" between consequent actions ("time of decision making"), differential equations (11a)–(11c) should be replaced by finite-difference equations. Although Eqs. (11a) and (11c) are deterministic equations, the system imitates random *trial-and-error* behavior if the latent time is not very small (**Figure 3**). It should be noted that such a pseudo-chaotic behavior of finite-difference equations' solution is quite typical for many nonlinear

[15]**Figure 1A**, rows 1 and 2 correspondingly.

[16]This is quite typical for the considered situation.

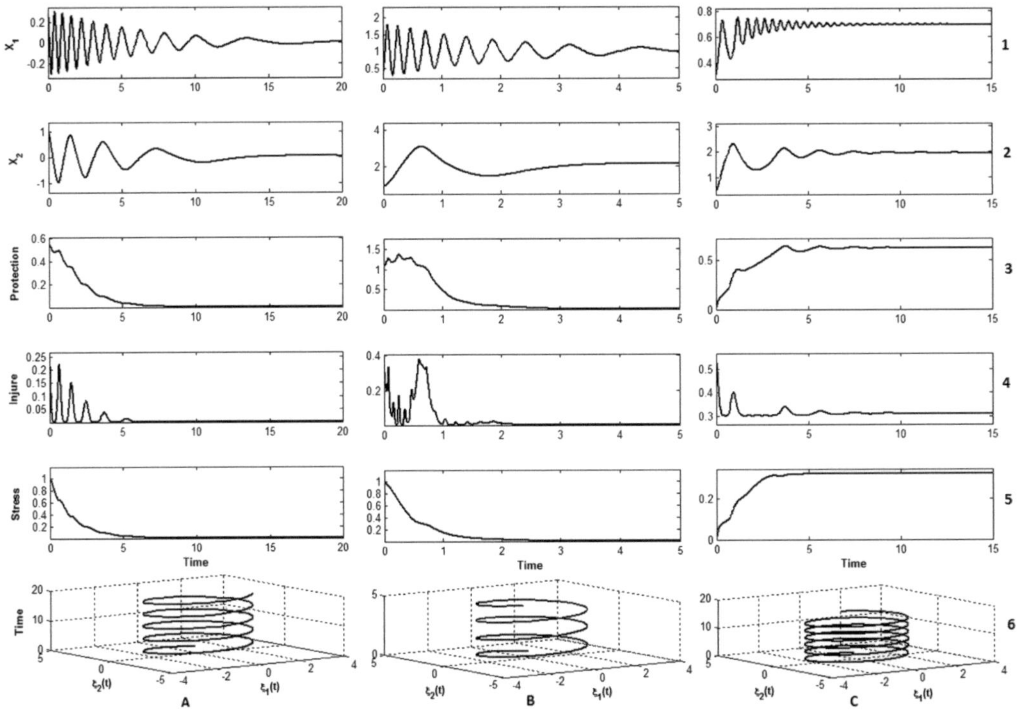

Figure 1. Homeostasis for different initial conditions. Here, x_1 and x_2 are *C*-variables and ξ_1 and ξ_2 are *R*-variables. (**A**) Light injury. (**B**) The system cannot return to the main ground state, but finds another comfortable state without damage and distress. (**C**) Homeostasis cannot fully compensate for injury and distress, but some discomforting stable state exists.

finite-difference equations and it was widely discussed in the literature. A particular example of such a behavior was considered in Ref. [17] and a general explanation of this phenomenon can be found in Ref. [19].

4. Discussion

Feeling of stress or proximity to death is a basic feature of the living organisms and this feature exists already at a single cell [1]. The discovery of homeostasis W. Cannon [3] assumed that homeostasis results from tendency of the organisms to decrease the stress and avoid death. This point is a biological basis for our theory.

It has been shown in Ref. [17] that if system evolution complies with the causality principle and a system state space displays trivial local topology, system dynamics inevitably satisfy generalized Lagrangian equations (2a) and (2b) with an additional "*S-variable.*" Since the above conditions are quite general, we believe that they are applicable to the living organisms. In the chapter, we identified *S-variable* with a level of feeling of stress (called *S-index*). It should be emphasized that the feeling of stress or discomfort is not metaphor for biological systems, but real feature of the living organisms (see Sections 1 and [1]). Note that *S-index* is a

http://dx.doi.org/10.5772/66473

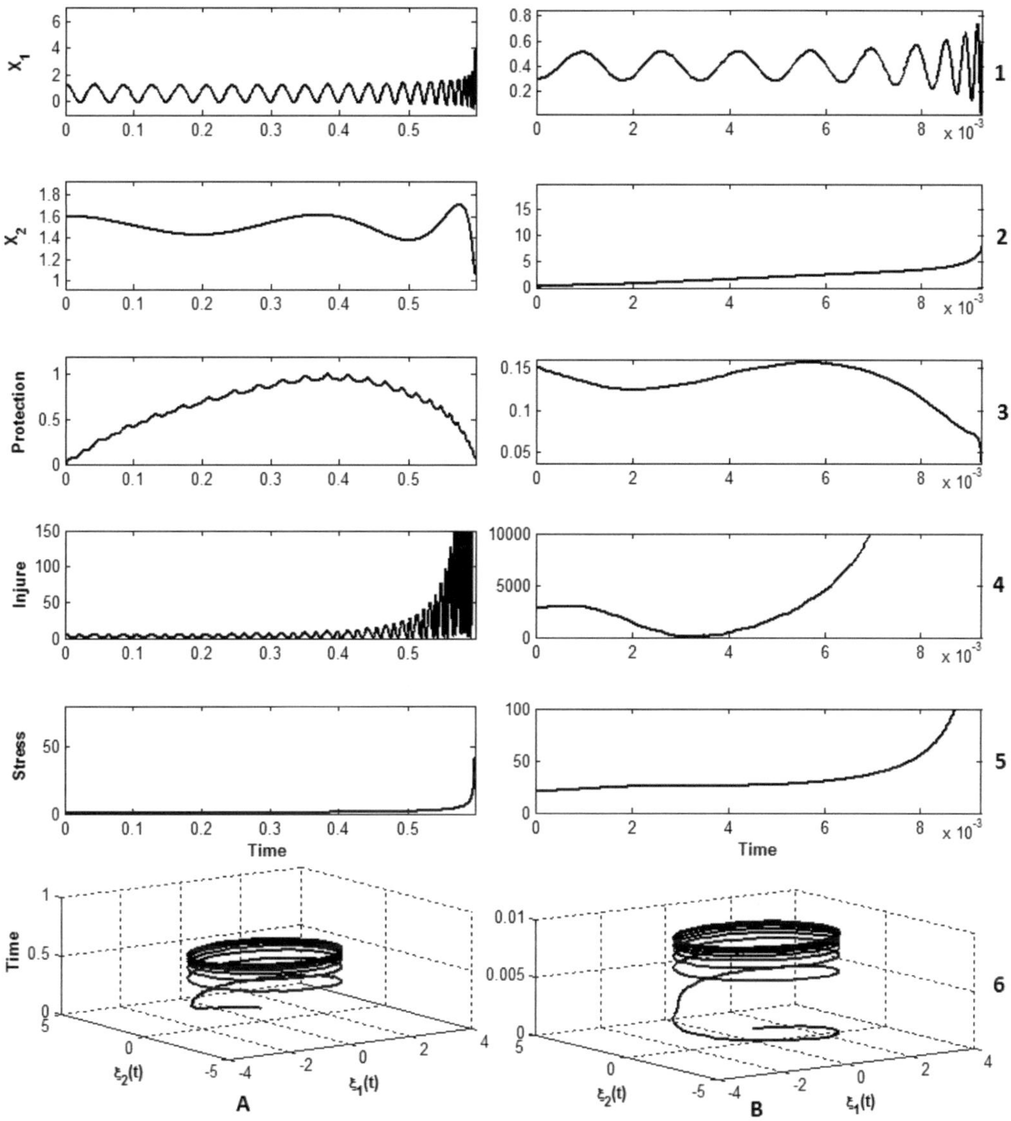

Figure 2. "Death-pathway" of the system. (**A**) System was heavily injured. (**B**) System was strongly distressed (initial *S-index* was high; graph **B5** begins from **20**), although initial injury was small.

phenomenological quantity and cannot be directly measured.[17] It should be noted that such a kind of the phenomenological variables (which cannot be directly measured) is widely used in physics (e.g., "mechanical action" of the physical systems, order parameter in superfluid phase transition, etc.).

Supposing that dynamics of the stress is determined by competition between damage and the protection mechanisms, we have obtained an *S-Lagrangian* and *dynamical equations of*

[17]Note, however, that in medical practice level of stress often is subjectively defined by the patients.

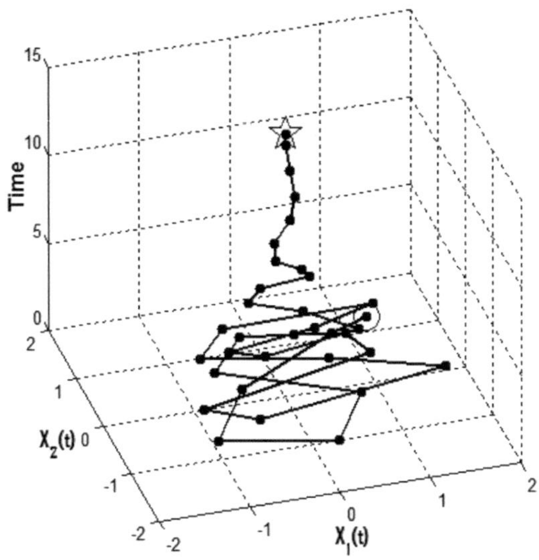

Figure 3. Pseudo-random behavior of the system with latent time of "decision making". *Circle* designates an initial state and *Star* designates the finish state.

homeostasis given in Eqs. (11a), (11b) and (11c). Moreover, since other systems, such as social systems, may also possess distress or discomfort, they may also undergo homeostasis.

Solutions of the dynamical equations of homeostasis show that there are four types of system behavior. In the first, the system generates activity that quickly takes it to the main ground state with zero damage and stress (**Figure 1A**). In the second, the main ground state cannot be achieved; however, the system finds another ground state without damage and stress as well (**Figure 1B**). In the third, homeostasis cannot find the state with zero damage and stress and the system arrives at the damaged and distressed, but stable stationary states (**Figure 1C**). In the last type of behavior, the system cannot achieve any stable state, level of stress dramatically increases, system variables leave the life-compatible region, and the system moves toward death (**Figure 2**). It should be noted that there is a critical value of injure, which leads to fatal instability of a system by violation of the condition (21). Apparently, there is a critical value of the stress as well, so if *S-index* exceeds this value, an organism inevitably moves toward death. Note that near the injured stable states, where $I(\boldsymbol{x}_c,S_c) > 0$, the critical value of the stress may be lower than near uninjured states,[18] that is, injured organism is more sensitive to the stress than the healthy one.

All types of behavior are described by the same system of Eqs. (11b) and (11c) and *S*-Lagrangian, but differ by initial and/or environmental conditions (which are described by parameters of the Lagrangian). It was found that systems exhibiting homeostasis may have at least two types of variables. The first type is *C-variables*, which have stationary values in the stable states of the system. Injury disturbs these values and excites protection mechanisms. The other types

[18]For $I(\boldsymbol{x},S)$ and $A(\boldsymbol{x},S)$ from Eqs. (29a) and (29b), the critical value of *S-index* is obtained from $\frac{\partial\Phi_1}{\partial S_*}J-(\Phi_2(S_*)+S_*\frac{\partial\Phi_2}{\partial S_*})\Gamma=0$.

http://dx.doi.org/10.5772/66473

of variables are *R-variables*, which can run in a stable state without disturbing system comfort. This finding agrees with the experimental data. Examples of the *C-variables* are ATP level, intracellular pH level, intracellular and blood concentration of Na^+, K^+, Ca^+, and intracellular levels of certain proteins (caspases, cytokines, and antioxidants), and so on. Examples of the *R-variables* are blood flow, α-rhythm of brain, heart contraction, brain pacemakers, and so on.

Interestingly, if a system has a "latent time" between consequent actions ("decision-making time"), it imitates random *trial-and-error* behavior. This corresponds to a real situation in a brain. Although the physical parameters of the brain are continuously changing, time intervals that are shorter than the nerve impulse duration (milliseconds) do not have physiological sense. Moreover, decisions in the brain take tens of milliseconds. Therefore, psychological time is more discrete. Consequently, the chaotic behavior of nerve processes inevitably arises in nerve tissue and can serve as the basis of free decision-making target. This creates an opportunity for trial-and-error behavior. A random search will be targeted if instability fluctuations increase with increasing deviation from the optimum. For example, on/off switching of voltage-dependent channels in neurons can occur more than 100 times/s, which is an adequate speed for searching for the homeostatic optimum. This mechanism ensures that obstacles can be overcome [20]. Therefore, chaotic behavior, illustrated in **Figure 3**, can play a crucial role in homeostasis. It should be emphasized that this chaotic behavior is not determined by some stochastic process, but rather is governed by deterministic equations.

The simulation results displayed satisfactory agreement between the biological properties of homeostasis and theory. **Figure 1** demonstrated direct homeostasis for a weak injury and indirect homeostasis with the restructuring of some parameters for more severe damage. Damage aggravation caused the model to transfer to a working state, although the discomfort was not completely removed. Modeling was also amenable to the process of system destruction (**Figure 2**), with the behavior of the model depending not only on damage severity but also on "subjective" assessment (i.e., death threats).

The theory predicts that increasing of the stress itself (even without internal injure) leads to disturbing of the physiological parameters, that is, to physiological damage of the organism. This prediction is supported by the recent experimental data, which show that both in human and in animal models, the expression of many genes changed in response to early and to late stresses [21].

Author details

Uziel Sandler[1]* and Lev Tsitolovsky[2]

*Address all correspondence to: sandler@g.jct.ac.il

1 Lev Academic Center (JCT), Jerusalem, Israel

2 Bar-Ilan University, Ramat Gan, Israel

References

[1] Sandler U., Tsitolovsky L., Neural Cell Behavior and Fuzzy Logic, Springer, NY (2008).

[2] Tsitolovsky L.E., Anticipation, consciousness, endogenous generation of goals and homeostasis, Int. J. Gen. Systems Spl. Iss. **44**, p. 655–666 (2015).

[3] Cannon W.B., Stresses and strains of homeostasis, Am. J. Med. Sci. **189**, p. 13–14 (1935).

[4] Doron G., von Heimendahl M., Schlattmann P., Houweling A.R., Brecht M., Spiking irregularity and frequency modulate the behavioral report of single-neuron stimulation, Neuron **81**, p. 653–663 (2014).

[5] Tognoli E., Kelso J.A.S., The metastable brain, Neuron **81**, p. 35–48 (2014).

[6] Bergquist S., Dickman D.K., Davis G.W., A hierarchy of cell intrinsic and target-derived homeostatic signaling neuron, Neuron **66**, p. 220–234 (2010).

[7] Thompson J.W., Bruick R.K., Protein degradation and iron homeostasis, BBA-Mol. Cell Res. **1823**, p. 1484–1490 (2012).

[8] Arianna Maffei, Alfredo Fontanini, Network homeostasis: a matter of coordination. Current Opinion in Neurobiology, Volume 19, Issue 2, 2009, Pages 168-173.

[9] Mehta A., Prabhakar M., Kumar P., Deshmukh R., Sharma P.L., Excitotoxicity: bridge to various triggers in neurodegenerative disorders, Eur. J. Pharmacol. **698**, p. 6–18 (2013).

[10] Wedesnky N.E., Excitation, Inhibition and Narcosis (In Russian), Publisher M.M. Stasiulevitch, St. Petersburg, 315 p.(1901).

[11] Sohl G., Maxeiner S., Willecke K., Expression and functions of neuronal gap junctions, Nat. Rev. Neurosci. **6**, p. 191–200 (2005).

[12] Wang H., Hu Y., Tsien J.Z., Molecular and systems mechanisms of memory consolidation and storage, Prog. Neurobiol. **79**, p. 123–135 (2006).

[13] Zadeh L.A., Fuzzy sets, Inform. Control **8**, p. 338–353 (1965).

[14] Zadeh L.A., Fuzzy sets as a basis for a theory of possibility, Fuzzy Set Syst. **1**, p. 3–28 (1978).

[15] Zadeh L.A., Possibility theory and soft data analysis, In: Cobb L., Thrall R.M. (Eds.), Mathematical Frontiers of the Social and Policy Sciences, Westview Press, Boulder, CO, p. 69–129 (1981).

[16] Zadeh L.A., Generalized theory of uncertainty (GTU) principal concepts and ideas, Comput. Stat. Data Anal. **51**, p. 5–46 (2006).

[17] Sandler U., Generalized Lagrangian dynamics of physical and non-physical systems, Phys. A **416**, p. 1–20 (2014). doi:10.1016/j.physa.2014.08.016.

[18] Boecker H., et al., The runner's high-opioidergic mechanisms in the human brain, Cerebral Cortex **18** (11), p. 2523–2531 (2008).

[19] Zaslavsky G.M., Sagdeev R.Z., Nonlinear Physics: Set Of, Harwood Academic, Newark, NJ (1991).

[20] Tsitolovsky L.E., Protection from neuronal damage evoked by a motivational excitation is a driving force of intentional actions, Brain Res. Rev. **49**, p. 566–594 (2005).

[21] Malki et al., The endogenous and reactive depression subtypes revisited: integrative animal and human studies implicate multiple distinct molecular mechanisms underlying major depressive disorder, BMC Med. **12**, p. 73–87 (2014).